2022 2nd International Conference on Electron Devices and Applications (ICEDA 2022)

Journal of Physics: Conference Series Volume 2331

Singapore
13-15 May 2022

ISBN: 978-1-7138-6050-1
ISSN: 1742-6588

Printed from e-media with permission by:

Curran Associates, Inc.
57 Morehouse Lane
Red Hook, NY 12571

Some format issues inherent in the e-media version may also appear in this print version.

Institute of Physics
Dirac House, Temple Back
Bristol BS1 6BE UK

Phone: 44 1 17 929 7481
Fax: 44 1 17 920 0979

techtracking@iop.org

Additional copies of this publication are available from:

Curran Associates, Inc.
57 Morehouse Lane
Red Hook, NY 12571 USA
Phone: 845-758-0400
Fax: 845-758-2633
Email: curran@proceedings.com
Web: www.proceedings.com

TABLE OF CONTENTS

Preface

Peer Review Statement

Graphene Bandgap Opening by Constructing Superlattices with BN or MoO_2 Under Pressure 1
Bohan Wang, Chaoyang Song

Numerical Investigation on Hole-Injection Characteristics of NiO/SiC Heterojunction 10
Wang Xi, Li Na, Pu Hongbin, Yang Yingxiang, Hu Jichao, Xu Jianning

Surface Roughness by Selective-Dissolving on PVA/PVP Nanofibers with Mobility Driven Concentration Polarization 15
Zheng Zhang, Feipeng Wang

Asphalt Interface Based on Abaqus Asphalt Pavement Temperature Field and Aggregate Morphological Characteristics Measurement System 24
Zhijun Ying, Huimin Wang, Yunfei Zhao, Zhihao Wu, Sili Shen

Investigation of 2D Material Anodes with Different Anions for Lithium Ion Batteries: Comparison of MoO_2, MoS_2 and $MoSe_2$ 38
Tianrui Chen

Investigation and Improvement of Switching Characteristics of SiC Optically Controlled Transistor 46
Yingying Zhang

Preparation and Investigation Indium Tin Oxide for Transparent Electrode of MS Schottky Photodiode 51
N Sangwaranatee, I Srithanachai

Effect of Irradiation (SRFE) on MSM Photodetectors Device 56
Itsara Srithanachai, Narong Sangwaranatee

Research on Hardware System Design Method of the Intelligent Vehicle Based on Dual Sensors 61
Wenliang Li

A BCF Bionic Robot Fish Driven by a Dielectric Elastomer Actuator 72
Zhi-Wei Jiao, Hao-Yu Wang, Bin Luo, Wei-Min Yang, Yuan Yu

Numerical Investigation on the Atomization Characteristics of an Aerosol Spray Nozzle 81
Liang Liu, Xingwei Wu, Dian Wang, Yue Yin

Design of a Novel Wireless NoC Architecture for Chip Multiprocessor 87
Yin Xin, Guozhi Song

The Design and Realization of Automobile Anti-Theft Control Circuit Based on SCM 96
Xue Li, Qisheng Li

Prediction Model of Elastic Load Resource Regulation Potential Under Hierarchical and Partitioned Dynamic Control Architecture 106
Ying Zhou, Songsong Chen, Kun Shi, Taorong Gong, Jindou Yuan, Bo Li, Shihai Yang

Author Index

Journal of Physics: Conference Series **2331** (2022) 011001 doi:10.1088/1742-6596/2331/1/011001

Preface

2022 2nd International Conference on Electron Devices and Applications was held virtually during May 13-15, 2022 via Zoom platform due to the pandemic of COVID-19. With a satisfactory ending of ICEDA 2021, ICEDA 2022 has been successfully held. Although the ravages of COVID-19 have prevented researchers and practitioners from sharing ideas in onsite way, benefiting from the development of information technology, all the practitioners can share what they thought and what they found via online platform. In this online platform, human ideas collide with spark.

ICEDA is designed to provide a platform for researchers and practitioners from electron devices and applications and related areas who are interested in developing, studying and using electron devices to share their research experiences and indulge in interactive discussions and communications at the event.

After strict review process, 18 papers were accepted and presented and included into conference proceedings. Many authors who come from different countries including China, Israel, Turkey, Egypt, Ireland, and Thailand proactively submitted their papers to conference committees. Based on the papers, all the researchers and practitioners can share the research fruits in electron devices. Also, attendees enthusiastically raised their questions which have been nicely answered by speakers. This conference gets unanimous praise from participants in that it facilitates the communication and dissemination of useful and insightful knowledge.

We are sincerely grateful to Advisory Committee Chair, Conference Chair, Conference Co-Chair, Program Chairs, Publication Chair, and International Technical Committees. Thanks for their generous support to ICEDA. Especially, we would like to be appreciated for technical committees and reviewers for taking time to review papers rigorously. We will try our best to make participants have fruitful discussions and share ideas freely in the platform of ICEDA.

Prof. Maode Ma
Qatar University, Qatar

Journal of Physics: Conference Series **2331** (2022) 011001 doi:10.1088/1742-6596/2331/1/011001

Conference Committee

Advisory Committee Chair

Xianbin Wang, Western University, Canada

Conference Chair

Maode Ma, Qatar University, Qatar

Conference Co-Chair

Massimo Alioto, National University of Singapore, Singapore

Program Chairs

Jeff Kilby, Auckland University of Technology, New Zealand
Jun Wu, Southeast University, China

Publication Chair

Letian Huang, University of Electronic Science and Technology of China, China

International Technical Committees

Prof. Tayeb Mohammed-Brahim, University of Rennes 1, France
Prof. Byung Seong Bae, Hoseo University, Korea
Prof. Razika Zair Tala-Ighil, University M'hamed Bougara Boumerdes, Algeria
Dr. Hieu Pham Trung Nguyen, University Heights, USA
Dr. Sujun Guan, Tokyo University of Science, Japan
Prof. Alexander M. Gaskov, Moscow State University, Russia
Prof. Ridvan Durak, Ataturk University, Turkey
Dr. Suraya Kamil, Universiti Teknologi MARA, Malaysia
Prof. Poi Sim Khiew, The University of Nottingham, Malaysia
Prof. Sotirios Baskoutas, University of Patras, Greece
Assoc. Prof. Sha Shiong Ng, Universiti Sains Malaysia, Malaysia
Prof. Ajay Kumar, Manipal University Jaipur, India
Prof. Duu Sheng Ong , Multimedia University, Malaysia
Dr. Mas Ayu Hassan, Universiti Malaysia Pahang, Malaysia
Dr. Shin Yee Khoo, University of Malaya, Malaysia
Dr. Ari Handono Ramelan, Sebelas Maret University, Indonesia
Dr. Rahul Davis, Sam Higginbottom University of Agriculture, Technology & Sciences (SHUATS), India
Dr. Ramadhansyah Putra Jaya, Universiti Teknologi Malaysia, Malaysia

Peer Review Statement

All papers published in this volume have been reviewed through processes administered by the Editors. Reviews were conducted by expert referees to the professional and scientific standards expected of a proceedings journal published by IOP Publishing Publishing.

- **Type of peer review:** Double Anonymous
- **Conference submission management system:** EasyChair
- **Number of submissions received:** 32
- **Number of submissions sent for review:** 32
- **Number of submissions accepted:** 18
- **Acceptance Rate (Submissions Accepted / Submissions Received × 100):** 56.3
- **Average number of reviews per paper:** 2
- **Total number of reviewers involved:** 17
- **Contact person for queries:**
  Name: Letian Huang
  Email: huanglt_uestc@outlook.com
  Affiliation: University of Electronic Science and Technology of China, China

Journal of Physics: Conference Series **2331** (2022) 012001 doi:10.1088/1742-6596/2331/1/012001

Graphene bandgap opening by constructing superlattices with BN or MoO_2 under pressure

Bohan Wang[1,*], Chaoyang Song[2]

[1]Beijing National Day School, China
[2]College of Chemistry and Molecular Engineering, Peking University, China

*Email: leo_wangbh@163.com

Abstract. After the isolation of the single layer graphene in 2004, which has overthrown the prediction of the two-dimensional material field, various aspects of its unique properties have been observed, including extremely high strength, carrier mobility and heat conductivity. Being the minority of nonmetallic materials with a conductive property as metals, its possible applications is much widened. Noteworthily, graphene has a zero bandgap. Although making it an excellent conductor, zero bandgap has also constrained its performance as a semiconductor. Fortunately, there are several possible ways to open its bandgap and make it behave like a semiconductor. In this study, we propose that constructing superlattices and controlling hydrostatic pressure to open the bandgap of graphene. Based on first-principles calculations, BN and MoO_2 are used to form the superlattice with graphene and 1-30 GPa pressures in the vertical direction of two-dimensional plane are applied. The results show that the pressure can affect the structures and interlayer distances of the superlattice, which further lead to a bandgap of the superlattices containing graphene. Our research shows a method that adjusting the bandgap of graphene through pressure, which is of great significance for the application of graphene in the electronic field.

1. Introduction

The two-dimensional material is defined as that the electrons can only distribute in the range of two dimensions and the out-plane thickness is generally lower than 10 nm [1]. In addition to graphene, two-dimensional materials such as transition metal dichalcogenides (TMDs) and MXenes have also been developed. Due to the wide variety chemical compositions and electron structures of two-dimensional materials, the properties are also very wide, including metals, semiconductors and insulators. In recent years, as the size of silicon-based field-effect transistors (FETs) has reached the theoretical limit, more and more researches have focused on the development of FETs based on two-dimensional materials [2-6]. So far, due to the excellent properties of two-dimensional materials, there are great prospects in the fields of electronics.

Graphene is the one layer unit of graphite, which was successfully separated by Geim et al. in 2004. As to the structure, graphene with a single atomic layer thickness of 0.34 nm, and is a two-dimensional layered single crystal composed of carbon. The carrier mobility of graphene is about 10000 $cm^2V^{-1}s^{-1}$ at room temperature[7-8], and thermal conductivity is 3000-5000 $Wm^{-1}K^{-1}$. Owing to the more excellent properties than traditional three-dimensional materials, it has broad application prospects in various fields. However, graphene's zero bandgap has constrained its application as a semiconducdor, where the valence and conduction bands meet at Dirac points, allowing electrons to move freely.

Therefore, although graphene has a high carrier mobility, the bandgap width is almost zero, which makes it difficult for graphene-based field effect transistors to have the switching state of the device. If modification methods such as doping or functionlization are used for the bandgap opening, these modifications would lead to a decrease in the conductivity of graphene. Therefore, the application of graphene in electronic devices such as next-generation field-effect transistors requires opening the bandgap and maintains the high carrier mobility [9].

Many researchers proposed methods to open the bandgap of graphene. For instance, Nemnes et al. [10] used HfO_2 substrates contacting with graphene, and when O atom is on the surface of HfO_2, the graphene can acquire a bandgap of ~1.8 eV. They attributed the bandgap opening to the exposed O atoms, which induce a doping effect to graphene. Ke et al. [11] experimentally studied the pressurized trilayer graphene, and found when the pressure larger than 30 GPa, the resistance increases exponentially, indicating a bandgap opening. Yankowitz et al. [12] constructed a heterostructure of graphene and boron nitride and applied pressure in the vertical direction to open the bandgap. They found that the bandgap is related to the value of pressure, and the bandgap is induced by the structure changing under pressures.

The 'zero bandgap' also means that it is sensitive to small change of external environments. With an electric field attraction produced by a polar material, such as molybdenum dioxide (MoO_2) or Boron Nitride (BN) attached to and pressurized on the single layer of graphene, the unbonded electron will be deviated, creating a bandgap. Therefore, in this study, we theoretically construct the graphene/MoO_2 and graphene/BN superlattices and calculate the electron band structures and charge density distributions of the superlattices under different pressures. We find the bandgap can be induced by the pressures when the pressure > 10 GPa, and bandgap > 1 eV when pressure reaches 30 GPa.

2. Methods

This study is based on first-principles calculations. All the simulations are modeled by Materials Studio software, and calculations are performed in the module of CASTEP [12]. Generalized gradient approximation (GGA) is used for the description of exchange correlation potential, and Perdew-Burke Ernzerhof (PBE) formulation is also applied [13]. A k-point mesh of $7\times7\times1$ and convergence criterion of 0.02 eV/Å are adopted for all simulations. Lastly, we introduce the DFT-D3 method for van der Waals interactions [14].

3. Results and discussions

Figure 1 shows the structures of graphene superlattices with pressures, where (a) pristine graphene (G), (b) BN, (c) MoO_2, and (d) graphene with *h*-BN (BN with hexagonal structure) superlattice (GBN) and (e) graphene with MoO_2 superlattice (GMO) under 1 GPa pressure in *z*-axis direction are displayed. The upper part is a top view, and the lower part is a front view. It can be seen that both graphene and boron nitride are with single-layer structures, and their atomic structures are very similar. MoO_2 has a three-layer structure, with O atoms in the upper and lower layers, and Mo atoms in the middle layer. The two layers of oxygen atoms have the same *x* and *y* coordinates. Due to graphene, boron nitride, and MoO_2 are all having hexagonal crystal structures, the shape changes little when heterostructure was formed.

In order to better show the effect of stress on the structure, the change of the lattice parameters of the heterostructure with pressure is calculated. Table 1 displays the lattice constant variations of G, BN, MoO_2 GBN and GMO under well relaxation and pressure, where 1 GPa, 5 GPa, 10 GPa, 20 GPa, and 30 GPa are considered. We do not show the structures of the heterostructures at 0 GPa pressure because in that case the interlayer distances are very large and the properties should be same with their monolayer materials. Lattice constants of G, BN and MoO_2 are 2.446, 2.512, 2.822 Å. After forming the heterostructure, the *a* or *b* lattice constant of the heterostructure is approximately equal to the average of the two materials. For example, GBN at 1GPa has lattice constant of *a* or *b* is 2.485 Å, with

2.485~(2.446+2.512)/2. And GMO at 1GPa has lattice constant of *a* or *b* is 2.667 Å, with 2.667~(2.446+2.822)/2. However, the *c* lattice constant shows *a* significant change with pressure. Lattice constants of *c* of GBN are 12.742, 7.904, 7.276, 6.900, 6.611 Å when imposing 1 GPa, 5 GPa, 10 GPa, 20 GPa, and 30 GPa pressures, and these of GMO are 10.080, 8.445, 7.917, 7.379 and 7.057 Å. This change in the layer spacing may cause changes in the 2D structure, which in turn transforms the electronic band structure of the superlattices.

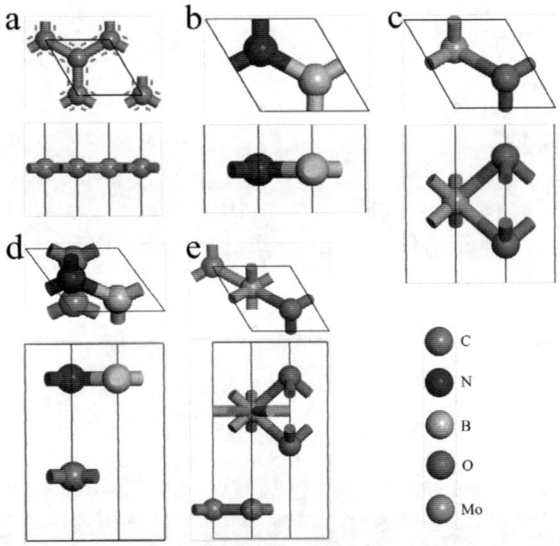

Figure 1. Structures of graphene superlattices, where (a) pristine graphene (G), (b) BN, (c) MoO_2, and (d) graphene with *h*-BN superlattice (GBN) and € graphene with MoO_2 superlattice (GMO) under 1 Gpa pressure in z-axis direction are displayed.

Table 1. Lattice constant variations of G, BN, MoO_2, GBN and GMO under well relaxation and pressure.

Materials	a(b)/Å	c/Å
G	2.446	----
BN	2.512	----
MoO_2	2.822	----
GBN-1GPa	2.485	12.742
GBN-5GPa	2.493	7.904
GBN-10GPa	2.490	7.276
GBN-20GPa	2.486	6.900
GBN-30GPa	2.483	6.611
GMO-1GPa	2.667	10.080
GMO-5GPa	2.668	8.445
GMO-10GPa	2.671	7.917
GMO-20GPa	2.677	7.379
GMO-30GPa	2.683	7.057

Figure 2 shows the band structures and density of states (DOSs) of (a) G, (b) BN, (c) MoO_2 monolayers, and (d) the high-symmetry points in the first Brillouin zone. It can be seen that the electronic band structure of graphene exhibits zero bandgap characteristics. At the Fermi level, the conduction band and the valence band are tangent at the high symmetry point K, and the density of states at the corresponding Fermi level is also almost zero. In addition, both BN and MoO_2 are insulators. The bandgaps of the former is about 5 eV, and of the latter is about 2 eV. If there is no interaction between the two layers, then the GBN and GMO superlattice should be the same as graphene, i.e. also a zero bandgap semiconductor.

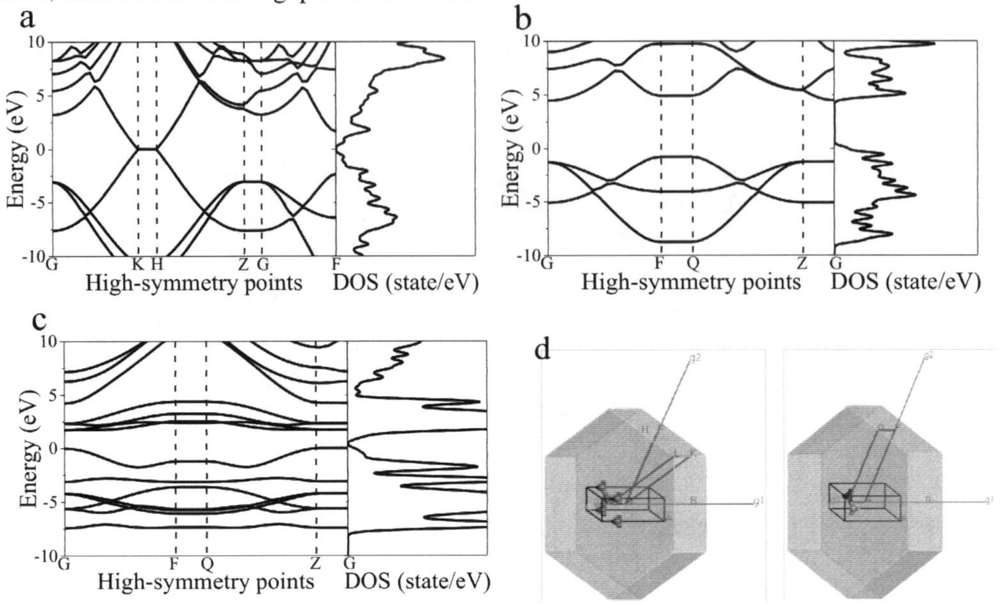

Figure 2. Band structures and density of states (DOSs) of (a) G, (b) BN, (c) MoO_2 monolayers, and (d) the high-symmetry points in the first Brillouin zone.

In order to study the electronic properties of superlattices and explore the influence of pressure, we calculated the electronic band structures of GBN and GMO superlattices under different pressures. Figure 3 shows the band structures and DOSs of GBN under pressure of (a) 1 GPa, (b) 5 GPa, (c) 10 GPa, (d) 20 GPa, (e) 30 GPa, and (f) is the highly symmetric paths and points in First Brillouin zone. It can be seen that when the pressure is small (<5 GPa), the electronic band structure near the Fermi level of GBN is similar to that of graphene. This is because BN hardly contributes electronic states near the Fermi level. Under low pressure, the interlayer interaction and structural changes of the superlattice are weak, so only the characteristics of graphene are shown near the Fermi level. However, as the pressure increases, the separation of the conduction band and the valence band near the K point can be found, which is different from the properties of graphene. The appearance of this bandgap may be caused by the reduction of the interlayer spacing caused by pressure.

Figure 4 shows the band structures and DOSs of GMO under pressure of (a) 1 GPa, (b) 5 GPa, (c) 10 GPa, (d) 20 GPa, (e) 30 GPa, and (f) is the highly symmetric paths and points in First Brillouin zone. The results of the electron band structures of GMO changing with pressure is similar to that of GBN. Under low pressure, it clearly displays that there is a tangent conduction band and a valence band at the Fermi energy, similar to graphene. And after the pressure is increased, the appearance of the bandgap can be captured, and the band gap increases as the pressure.

Figure 5 summarizes the bandgaps of G, GBN and GMO *vs*. pressure. The results of the band gap of GBN and GMO are acquired from Figs. 3 and 4, and the band structures of G under different pressures are not shown here because the properties of pure graphene has been well investigated First of all, pure graphene keeps 0 bandgap when the pressure is increased from 1 to 30 GPa, indicating that the pressure cannot make the graphene open the bandgap. Secondly, under a lower pressure of 1 GPa, the two superlattice materials GBN and GMO also maintain a bandgap of 0, indicating that only the forming superlattice also cannot open the bandgap of graphene. However, applying pressure to the superlattice material can cause the graphene bandgap to open at high pressure. The maximum bandgaps of GBN and GMO are respectively about 1.0 and 1.3 eV, which can basically meet the application of field effect transistors.

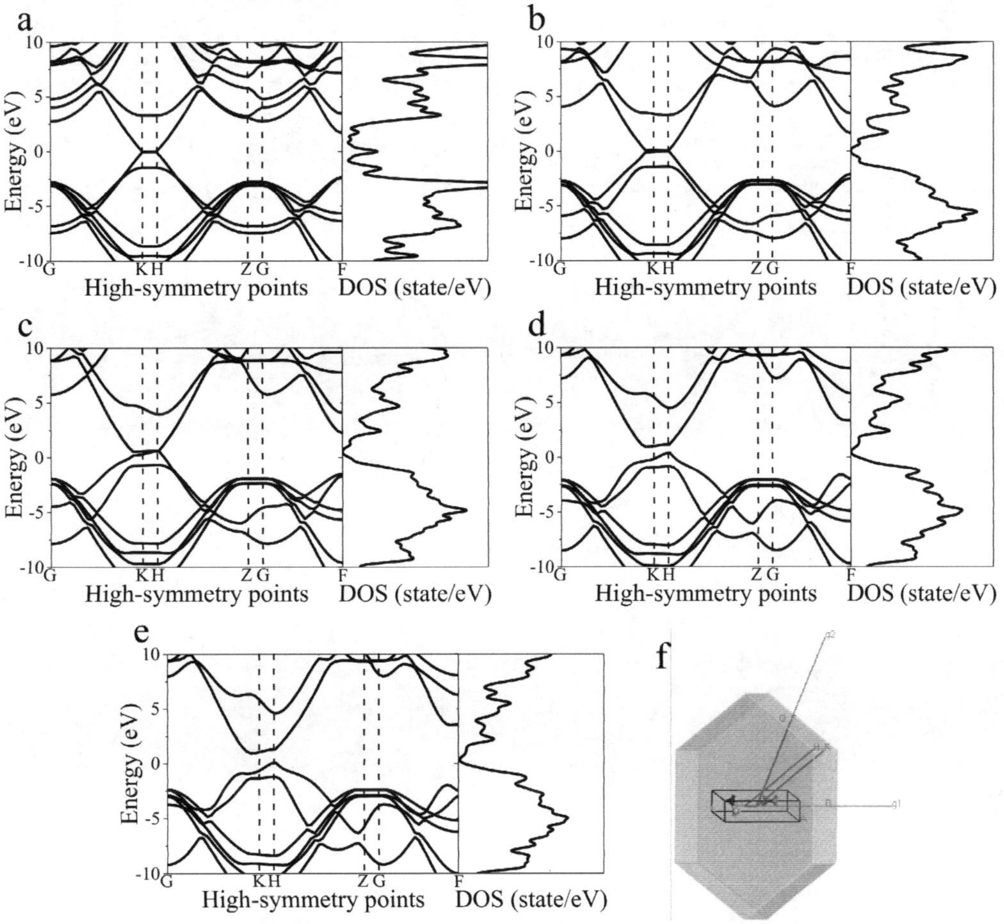

Figure 3. Band structures and DOSs of GBN under pressure of (a) 1 GPa, (b) 5 Gpa, (c) 10 Gpa, (d) 20 Gpa, € 30 Gpa, and (f) is the highly symmetric paths and points in First Brillouin zone.

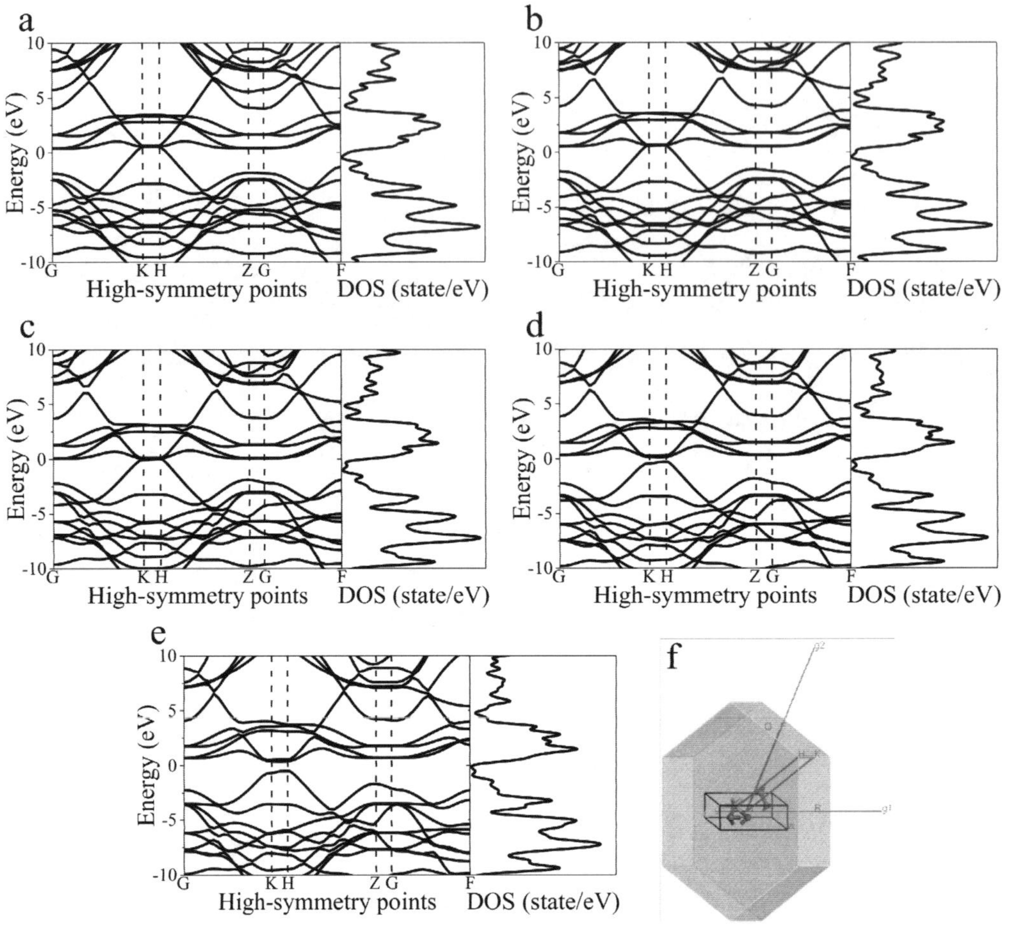

Figure 4. Band structures and DOSs of GMO under pressure of (a) 1 Gpa, (b) 5 Gpa, (c) 10 Gpa, (d) 20 Gpa, € 30 Gpa, and (f) is the highly symmetric paths and points in First Brillouin zone.

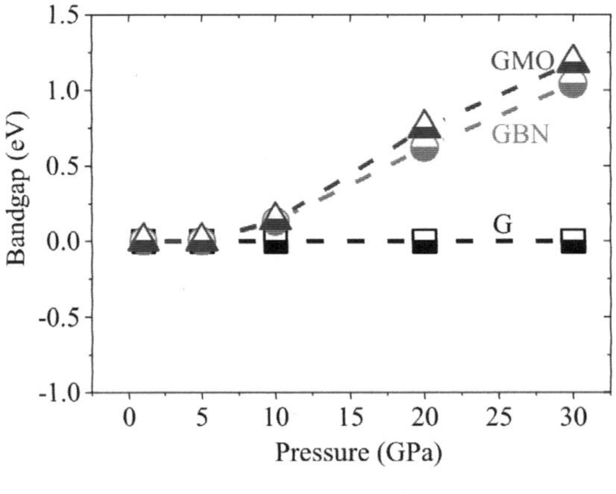

Figure 5. Bandgaps of G, GBN and GMO changing with pressure.

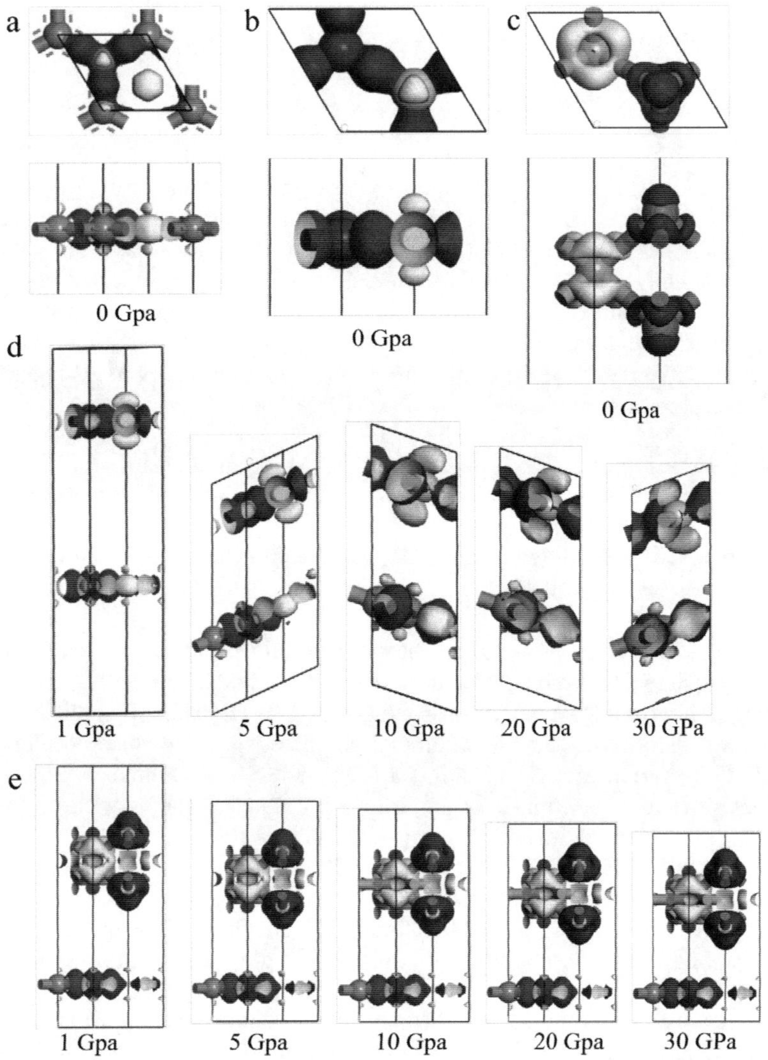

Figure 6. Electron density difference plots for (a) G, (b) BN, (c) MoO_2, (d) GBN and (e) GMO under well relaxation and pressure, where the blue and yellow isosurfaces represent charge accumulation and consumption, respectively.

In order to further discuss the reasons for the bandgap, we calculated the electron density difference of the three materials under various pressures. Figure 6 shows the electron density difference plots for (a) G, (b) BN, (c) MoO_2, (d) GBN and (e) GMO under well relaxation and pressure, where the blue and yellow isosurfaces represent charge accumulation and consumption, respectively. First of all, the charge of the C atoms in graphene gathers towards the position of the C-C bond, and the center of the six-membered ring of C exhibits charge consumption. The charge in BN is also concentrated near the B-N bond. At the same time, N atoms appear as charge accumulation, and B atoms appear as charge consumption, indicating that N has taken away the electrons of B atoms. The electrons in MoO_2 reflect the nature of ionic bonds, that is, the electrons of Mo atoms are completely taken away by O, and there is no charge accumulation on the Mo-O bond.

For GBN and GMO superlattices, the increase in pressure greatly reduces the interlayer spacing. Under high pressure, such as 30 GPa and in GBN, the charges of BN and graphene are very close, which may cause the charge transfer between the two 2D materials. At the same time, the strong charge repulsive force may also change the structure of 2D materials. The results of GMO and GBN can be understood similarly. This charge transfer and structural change may jointly lead to the opening of the bandgap of graphene.

4. Conclusion
Aiming at the issue that graphene has a band gap of 0 and cannot be used in field-effect transistors, first-principles calculations are performed to study the electronic band structure changes of graphene when forming a superlattice and under pressures. The results show that pure graphene has no bandgap under various pressures, and the superlattices GMO and GBN also have zero bandgap when the pressure is very low. Only when a higher pressure is applied to the superlattices, the bandgap will appear. The results of the charge density difference suggest that the opening of the graphene bandgap may be caused by the charge transfer between the two layers of 2D materials and the structure changes in each layer. Our study can guide the application of graphene in the field of electronics.

References
[1] Bonaccorso, F., Colombo, L., Yu, G., Stoller, M., & Pellegrini, V. (2015). 2d materials. graphene, related two-dimensional crystals, and hybrid systems for energy conversion and storage. Science, 347(6217), 1246501.
[2] Sedki, M., Chen, Y., & Mulchandani, A. (2020). Non-carbon 2d materials-based field-effect transistor biosensors: recent advances, challenges, and future perspectives. Sensors, 20(17), 4811.
[3] Gao, F., Yang, H., & Hu, P. A. (2018). Interfacial engineering for fabricating high-performance field-effect transistors based on 2d materials. Small Methods, 1700384.
[4] Wai, Leong, Chow, Peng, Yu, & Fucai, et al. (2017). High mobility 2d palladium diselenide field-effect transistors with tunable ambipolar characteristics. Advanced Materials, 29: 1602969.
[5] Marin, E. G., Marian, D., Perucchini, M., Fiori, G., & Iannaccone, G. (2020). Lateral heterostructure field-effect transistors based on 2d-material stacks with varying thickness and energy filtering source. ACS Nano 14, 2, 1982–1989.
[6] Shi, S., Hu, R., Wu, E., Li, Q., & Liu, J. (2018). Highly-sensitive gas sensor based on two-dimensional material field effect transistor. Nanotechnology, 29(43), 435502.
[7] K.I. Bolotin, K.J. Sikes, Z. Jiang, M. Klima, G. Fudenberg, J. Hone, P. Kim, H.L. Stormer, (2008). Ultrahigh electron mobility in suspended graphene. Solid State Communications, 146(9–10), 351-355.
[8] Wang, Q. H., Jin, Z., Kim, K. K., Hilmer, A. J., Paulus, G., & Shih, C. J, et al. (2012). Understanding and controlling the substrate effect on graphene electron-transfer chemistry via reactivity imprint lithography. Nature Chemistry, 4(9), 724.
[9] Meric, I., Han, M. Y., Young, A. F., Ozyilmaz, B., Kim, P., & Shepard, K. L. (2008). Current saturation in zero-bandgap, top-gated graphene field-effect transistors. Nature Nanotechnology, 3(11), 654-9.
[10] Nemnes, G. A., Dragoman, D., & Dragoman, M. (2019). Graphene bandgap induced by ferroelectric pca2 1 hfo 2 substrate: a first-principles study. Physical Chemistry Chemical Physics, 21(27), 15001.
[11] Ke, F., Y Chen, Yin, K., Yan, J., & Chen, B. (2019). Large bandgap of pressurized trilayer graphene. Proceedings of the National Academy of Sciences, 116(19), 201820890.
[12] Yankowitz, M., Jung, J., Laksono, E., Leconte, N., Chittari, B. L., & Watanabe, K., et al. (2018). Dynamic band-structure tuning of graphene moiré superlattices with pressure. Nature, 557(7705), 404-408.
[13] Clark, Stewart J., Segall, Matthew D., Pickard, Chris J., Hasnip, Phil J., Probert, Matt I. J., Refson, Keith and Payne, Mike C. "First principles methods using CASTEP" Zeitschrift für Kristallographie - Crystalline Materials, vol. 220, no. 5-6, 2005, pp. 567-570. https://doi.org/10.1524/zkri.220.5.567.65075

[14] J. P. Perdew, K. Burke, and M. Ernzerhof, Generalized gradient approximation made simple, Phys. Rev. Lett. 77, 3865 (1996).

Journal of Physics: Conference Series **2331** (2022) 012002 doi:10.1088/1742-6596/2331/1/012002

Numerical Investigation on Hole-Injection Characteristics of NiO/SiC Heterojunction

Wang Xi[1,*], Li Na[1], Pu Hongbin[1,2], Yang Yingxiang[2], Hu Jichao[1], Xu Jianning[1]

[1] Department of electronic engineering, Xi'an University of Technology, Xi'an, China
[2] Xi'an Key Laboratory of Power Electronic Devices and High Efficiency Power Conversion, Xi'an, China

* wangxii@xaut.edu.cn

Abstract. Numerical investigation on hole-injection characteristics of NiO/SiC heterojunction is carried out in this paper. Theory analysis and numerical simulation both indicate the excellent hole-injection characteristic of p-NiO/n-SiC heterojunction. The pn junction diode and pnp phototransistor are constructed and simulated to evaluate hole-injection characteristics p-NiO/n-SiC heterojunction. The results indicate that the p-NiO/n-SiC heterojunction shows great potential advantage in enhancing current gain of pnp phototransistor. By using NiO/SiC heterojunction as the emitter junction, the current gain of SiC based pnp phototransistor can be increased by about 96.3 times.

1. Introduction

As a wide band gap semiconductor material with direct band gap, NiO usually behaves as p-type semiconductor material with a wide band gap about 3.6 eV to 4.0 eV, and has been widely used as p-type region of heterojunction in photoelectric sensor, photodetector, solar cell and semiconductor power device [1-4]. Hehe Gong et. al. demonstrated a high power diode with NiO/Ga_2O_3 heterojunction, which indicated the promising potential of NiO in semiconductor power devices [5]. Jun-Dar Hwang et. al. reported a high performance bipolar phototransistor with NiO/ZnO heterojunction, which evaluated the advantage of NiO material in photodectors [6]. Chenlu Wang et. al. proposed a field-effect-transistor with based on NiO/Ga_2O_3 heterojunction, which shows the promises of NiO in boosting the Ga_2O_3 power device performances [7]. NiO has shown promising potential in above applications. However, reports on application of NiO in SiC based devices are very few [8, 9]. If the potential application advantage of NiO in SiC based devices is excavated, the application scope and role of NiO will be further expanded.

In this work, NiO/SiC heterojunction is investigated through numerical simulation. Hole-injection characteristics of NiO/SiC heterojunction are compared with SiC pn junction. Potential advantage of NiO/SiC heterojunction in SiC bipolar device is also demonstrated through SiC based phototransistor.

2. Theory analysis

The pn junction is a main structure of semiconductor devices. Injection characteristic of pn junction usually plays a key role on both static and dynamic performances of a semiconductor bipolar device. When forward biased, hole injection capacity of pn junction can be expressed by the ratio of hole current density to electron current density [10].

$$\frac{J_p}{J_n} = \frac{D_p p_{po} L_n}{D_n n_{no} L_p} \exp\left(\frac{\Delta E_C - \Delta E_V}{kT}\right) \quad (1)$$

Where D_p and D_n are diffusion coefficient of hole and electron, respectively. The p_{p0} and n_{n0} are concentrations of majority carriers in p region and n region, respectively. The L_n and L_p are diffusion length of electrons and holes, respectively. The $\triangle E_C$ and $\triangle E_V$ are conduction energy band discontinuity and valence energy band discontinuity, respectively.

Energy band diagrams of p-NiO/n-SiC heterojunction and SiC pn junction are shown in Fig. 1(a) and (b). For SiC pn junction, both conduction energy band difference and valence energy band difference are zero. Base on equation (1), the hole-injection ratio of SiC pn junction can then be expressed as,

$$\frac{J_p}{J_n} = \frac{D_p p_{po} L_n}{D_n n_{no} L_p} \quad (2)$$

However, because of difference materials, there are conduction energy band difference and valence energy band difference between NiO and SiC. Based on equation (1), the hole-injection ratio of p-NiO/n-SiC heterojunction can then be expressed as,

$$\frac{J_p}{J_n} = \frac{D_p p_{po} L_n}{D_n n_{no} L_p} \exp\left(\frac{\Delta E}{kT}\right) \quad (3)$$

As shown in Fig. 1(a), conduction energy band difference and valence energy band difference between NiO and SiC are 1.64 eV and 1.15 eV, respectively. The $\triangle E$ are calculated to be 0.31 eV. Therefore, p-NiO/n-SiC can obtain much higher hole-injection ratio than SiC pn junction in theory. On condition that ignoring diffusion coefficient and diffusion length between NiO and SiC, hole-injection ratio can increased up by exp(0.49/kT) times with same majority carrier concentration in theory.

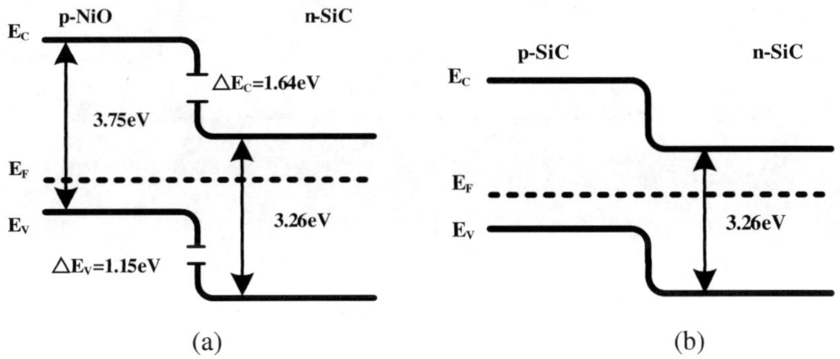

Figure 1. Energy band diagrams of pn junctions. (a) p-NiO/n-SiC heterojunction. (b) SiC pn junction.

3. Simulation and Discussion

3.1. NiO/SiC Heterojunction

Though p-NiO/n-SiC heterojunction exhibits enormous potential in improving hole-injection capacity of SiC pn junction, single crystal p-NiO/n-SiC has not been manufactured until now, result from immaturity material fabrication technology. In order to predict influence of p-NiO/n-SiC heterojunction on performances of SiC based bipolar devices, hole-injection characteristics p-NiO/n-SiC heterojunction of numerical simulation is carried out. Structure diagrams of p-NiO/n-SiC heterojunction and SiC pn junction diodes are shown in Fig. 2(a) and (b). In p-NiO/n-SiC heterojunction diode, thickness of p emitter region and n base region are 0.5 μm and 1.0 μm, respectively. Doping concentration of p emitter region and n base region are 1.0×10^{18} cm^{-3} and 1.0×10^{16} cm^{-3}, respectively. Except for material of p emitter region, parameters in both p-NiO/n-SiC heterojunction diode and SiC pn junction diode in this work are same with each other.

Fig. 3 shows forward current-voltage curves of p-NiO/n-SiC heterojunction diode and SiC pn junction diode. Hole concentrations of p emitter region in SiC pn junction diode are adjusted to change hole-injection characteristics of SiC pn junction diode in the simulation, through enhancing doping concentration and shielding incomplete ionization model. When shielding the incomplete ionization model, the acceptors in p emitter region are completely ionized and the hole concentration is increased. As shown in the figure, SiC pn junction diode shows lower on-resistance than p-NiO/n-SiC heterojunction diode. It is because that barrier for carriers in p-NiO/n-SiC heterojunction is combined with energy band bending and energy band difference. Corresponding barrier in SiC pn junction is only consisted by energy band bending. When increasing doping concentration of p emitter, SiC pn junction diode shows lower resistance. When shielding incomplete ionization model, resistance of SiC pn junction diode shows further reduction. It is because that majority carrier concentration can be increased by both enhancing doping concentration and shielding incomplete ionization model. According to equation (2), enhancement of majority carrier concentration in p emitter would enhance hole-injection ratio of SiC pn junction. As a result, stronger conductivity modulation is formed in n base region and the resistance is reduced. Because of the energy band mismatch between NiO and SiC, barrier height for carriers in p-NiO/n-SiC heterojunction is higher than that in SiC pn junction. Therefore, the forward current-voltage curves can not express the advantages of hole-injection characteristic of p-NiO/n-SiC heterojunction directly.

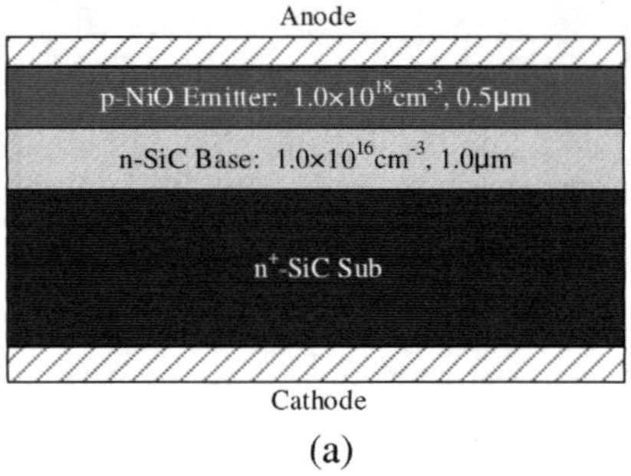

Figure 2. Structure diagrams of pn junction diodes. (a) p-NiO/n-SiC heterojunction diode. (b) SiC pn junction diode.

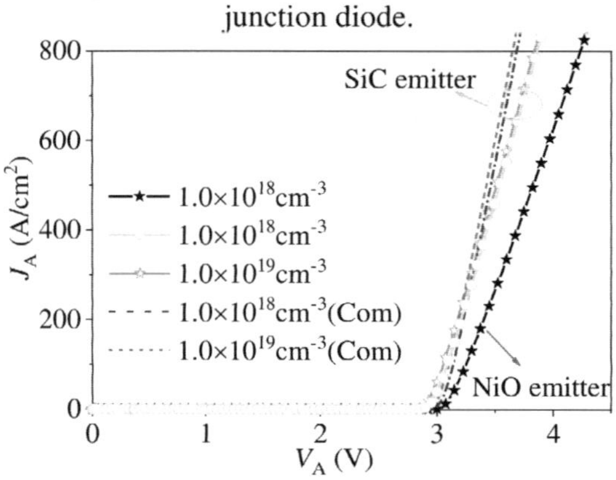

Figure 3. Forward current voltage curves of pn junctions.

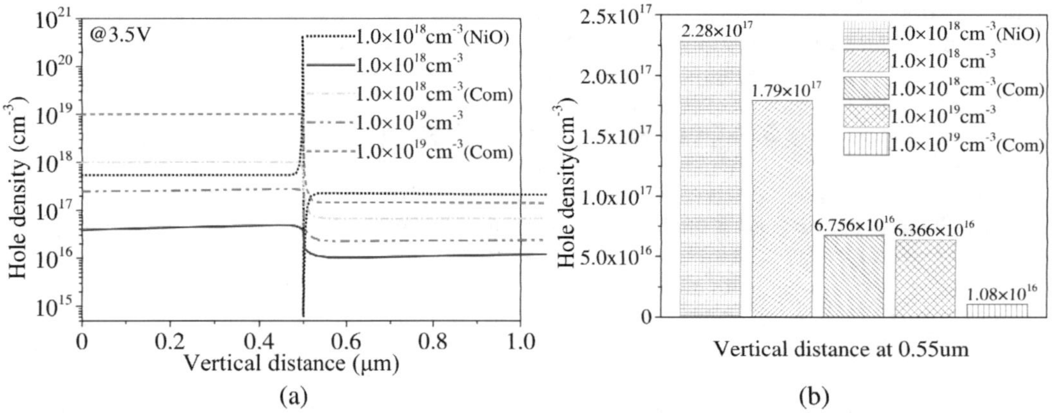

Figure 4. Hole densities in pn junctions at 3.5V forward bias. (a) Hole density distributions in pn junctions. (b) Peak value of hole densities in n region.

In order to further analyze hole-injection characteristic of p-NiO/n-SiC heterojunction, on-state hole density distributions in pn diodes are extracted and discussed. Fig. 4 shows on-state hole density distributions in p-NiO/n-SiC heterojunction diode and SiC pn junction diode at 3.5 V forward bias. As shown in the figure, SiC pn diode with higher hole concentration in p emitter region injects more holes into n base region and obtains higher hole concentration in n base region, which is in keeping with equation (2). However, profit from energy band difference between NiO and SiC, n base region of p-NiO/n-SiC heterojunction diode still obtains the highest hole density among these diodes, even though hole concentration in p NiO emitter is not the highest. The hole density distributions indicated the p-NiO/n-SiC heterojunction has stronger hole injection capacity than SiC pn junction, which is in keeping with equation (3).

3.2. NiO/SiC phototransistor

In order to investigate potential advantage of hole-injection characteristics of NiO/SiC heterojunction in SiC bipolar device, SiC based phototransistor with NiO emitter is carried out and simulated. Structure diagram of NiO/SiC heterojunction phototransistor is shown in Fig. 5(a). The thickness of n base region and p collector region in the NiO/SiC heterojunction phototransistor are 1.0 μm and 2.0 μm, respectively. The doping concentration of n base region and p collector region in the NiO/SiC heterojunction phototransistor are 1.0×10^{16} cm^{-3} and 1.0×10^{14} cm^{-3}, respectively. The doping concentration and junction depth of p heavily doped region under collector electrode are 1.0×10^{14} cm^{-3} and 0.5 μm, respectively. Fig. 5(b) shows the SiC pnp phototransistor. Except for material of p emitter region, parameters in NiO/SiC heterojunction phototransistor and SiC pnp phototransistor in this work are same with each other.

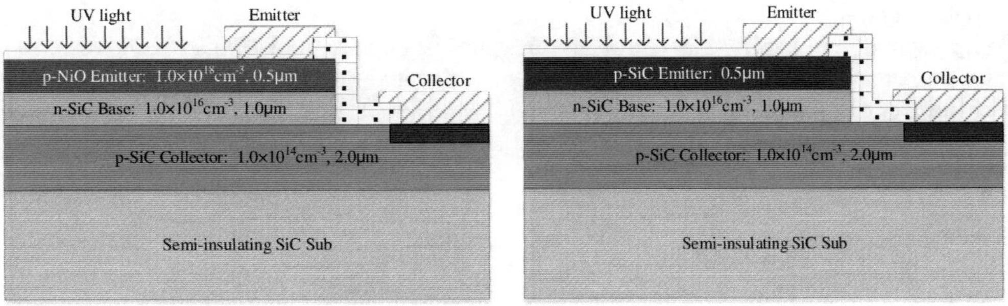

(a) (b)

Figure 5. Structure diagrams of pnp phototransistors. (a) NiO/SiC heterojunction phototransistor. (b) SiC pnp phototransistor.

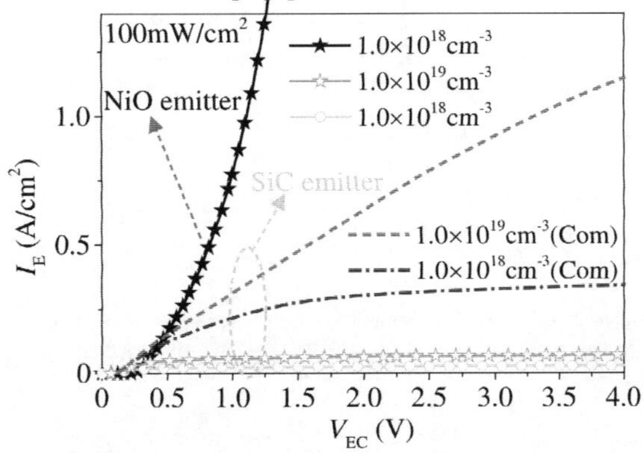

Figure 6. Output characteristic curves of pnp phototransistors.

The output characteristic curves of NiO/SiC heterojunction phototransistor and SiC pnp phototransistor are shown in Fig. 6. Consistent with pn junction diode, current gain in SiC pnp phototransistor is also increased with increasing of doping concentration of p emitter region. When shielding the incomplete ionization model, current gain of the SiC pnp transistor is further increased.

Even though, the NiO/SiC heterojunction phototransistor still obtains a much higher current gain than SiC pnp transistors. When emitter voltage bias is 1.5 V, current gain of the NiO/SiC heterojunction phototransistor is 23.4 A/W. The current gain in SiC pnp transistor with identical structure parameters is only 243 mA/W. The current gain of phototransistor can be increased by about 96.3 times by using NiO/SiC heterojunction. Despite shielding the incomplete ionization model in SiC pnp transistor, the current gain of phototransistor can be increased by about 49.3 times by using NiO/SiC heterojunction. The simulation results illustrate the importance of hole-injection characteristic on output characteristic of the transistor. Benefit from unique energy band structure, NiO/SiC heterojunction exhibits excellent hole-injection characteristic and shows big improvements on characteristics of pnp phototransistor.

4. Conclusion

Hole-injection characteristics of NiO/SiC heterojunction is investigated numerically in this paper. Energy band structures of p-NiO/n-SiC heterojunction and SiC pn junction are constructed and analyzed. The unique energy band structure of NiO/SiC heterojunction exhibits excellent hole-injection characteristic. Hole-injection characteristics p-NiO/n-SiC heterojunction are evaluated through numerical simulation of pn junction diode and pnp phototransistor. The results indicate that, despite its excellent hole-injection characteristic, p-NiO/n-SiC heterojunction is difficult to improve the turn-on performance of pn diode because of the higher barrier for carriers. However, p-NiO/n-SiC heterojunction shows great potential advantage in enhancing current gain of pnp phototransistor. The simulation results show that the current gain of pnp phototransistor can be increased by about 96.3 times by using NiO/SiC heterojunction.

Acknowledgments

This paper supported in part by National Natural Science Foundation of China (62004161), in part by Scientific Research Project of Education Department of Shaanxi Province (20JK0796), in part by Youth talent lift project of Xi'an Science and Technology Association (095920201318), in part by Shaanxi Innovation Capability Support Project (2022TD-39), and in part by Natural Science Basic Research Plan in Shaanxi Province of China (2020JQ-636).

References

[1] Li L, Tong Z, Ao J P. 2016, Surfaces & Interfaces, 5:15-18
[2] Zhang Z, Ning Y, Fang X. 2019, Journal of Materials Chemistry C, 7(2): 223
[3] Madani Labed, Nouredine Sengouga,et al. 2021, Optical Materials, 120: 111453
[4] Kokubun Y, Kubo S, Nakagomi S. 2016, Applied Physics Express, 9(9):091101
[5] H. Gong et al., 2021, IEEE Transactions on Power Electronics, 36(11):12213
[6] Hwang J D, Hwang Y E, Chen Y A . 2021, Nanotechnology, 32(40):405501
[7] Wang C, H Gong, Lei W, et al. 2021, IEEE Electron Device Letters, 42(4):485
[8] Chen C, Pu H, Wang M, et al. 2020, Physics Letters A, 384(33):126824
[9] Cho Seulki, Seo Ji-Ho, Lee Young-Jae, et al. 2018, Nanoscience and Nanotechnology Letters, 10(12):1700
[10] Zhiming C, Tianmin L, Jianping M. 2011, A Concise Course of Semiconductor Physics (Beijing: Machinery Industry Press)

Journal of Physics: Conference Series **2331** (2022) 012003 doi:10.1088/1742-6596/2331/1/012003

Surface roughness by selective-dissolving on PVA/PVP nanofibers with mobility driven concentration polarization

Zheng Zhang[1], Feipeng Wang[1*]

[1]School of Electrical Engineering, Chongqing university, No.174, Shazheng Road, Chongqing City, China 400030

fpwang@cqu.edu.cn

Abstract. Surface roughened polyvinyl alcohol (PVA) nanofibers are produced by electrospinning and selective dissolution in ethanol. By electrospinning of PVA (Mw 74885) and polyvinylpyrrolidone(PVP) (Mw 44000-54000) blend solution, PVA/PVP blend nanofibers with smooth, homogenous surface are prepared. Then the surface roughened nanofibers are achieved via selective dissolution in ethanol. Fourier Transform Infrared Spectra (FTIR) and Energy Dispersive Spectrometer (EDS) result shows a complete removal of PVP from blend nanofibers. In this work, PVA/PVP is the first water-soluble polymer combination adopted to create surface roughened nanofibers by selective dissolution. Moreover, opposite to other selective dissolution schemes which removes polymers of higher Mw, this is the first time creating surface roughened nanofibers by selectively removing polymer of lower Mw successively. Through selective dissolution, rough structures appear on the ultra-fine PVA-PVP nanofibers surface. The successive preparation of surface roughened PVA nanofiber can be attributed to the mobility of PVP molecule. The method claimed in this paper is helpful for developing more types of surface roughened nanofibers in the future.

1. Introduction

Nanofiber is a kind of fiber whose diameter is in nanometer scale. It now gets wide applications in air and water filtration, catalyst substrate[1], sensor[2], biomedical scaffold[3, 4] as well as drug carrier[5, 6]. One of the most prevalent methods to produce nanofibers is electrospinning. By applying electric field on the needle and the receiver, droplets are stretched under the synergy of electrostatic force and surface tension to form a jet. The nanofibers are obtained when jets fly to collectors with a process of solvent evaporation[7]. Though the size of nanofiber is limited to nanoscale, its type and structure are quite numerous, for example core-shell nanofiber[8], hallow nanofiber[9], tri-axial nanofiber[10], tree-like nanofiber[11], surface roughened nanofiber and so on. Among all the special structures of nanofiber, the surface roughened nanofibers have a lot of unique properties which makes them be applied in field of biological, sensors, construction and so on.

Selective dissolution is a method to remove certain component from blends by utilizing solubility difference of different compositions in a given solvent. Generally speaking, selective dissolution usually includes procedures to make surface roughened nanofibers as follows: At first, solutions including two polymers are prepared and electrospinning process is carried out to obtain blend nanofibers. Then a specific solvent in which one polymer could entirely dissolve and the other can not dissolve at all is used to remove one kind of polymer (sacrificial polymer) from the blend nanofibers. The nanofibers going through selective dissolution process show rough surface. A few polymer combinations have been adopted to make solutions and achieve surface roughened nanofiber, for example, polyethylene oxide (PEO) and polymethyl methacrylate (PMMA)[12], polycaprolactones (PCL) and PVP[13], PVA and polyacrylonitrile (PAN)[14], polyglycolideacid (PGA) and polylactic acid (PLA)[15], PAN and polystyrene (PS)[16]. However, as far as all the polymer combinations we can find now, the sacrificial polymers are the higher-Mw polymers rather than the lower-Mw ones in every polymer combinations, which limits the types of surface roughened nanofibers we can produce. The reason is that higher-Mw polymers distribute at the sheath of blend nanofibers and lower Mw polymers distribute at the core of blend nanofibers That is because the viscosity of polymer solution increases with Mw of polymer, which makes the the solidification of higher-Mw polymers become fast on nanofiber sheath[17]. It is especially difficult to obtain surface roughened nanofiber of high Mw polymers considering the difficulties to find a higher Mw one as sacrificial polymer and suitable solvent to selectively dissolve.

In this work, trials have been made to obtain the surface roughened PVA nanofibers. As far as we know, this is the first time combining two kinds of water soluble polymers and selectively dissolving one of them to produce surface roughened nanofibers. The molecular structure of PVA and PVP is exhibited in Figure 1. Furthermore, this is also the first time surface roughened nanofiber is created by removing lower-Mw polymer composition from bio-component blends, which provides practical experience for ongoing research about nanofiber surface modification. FTIR testing and EDS are adopted to make sure the complete removal of PVP. Morphology of blend and surface roughened nanofibers has been recorded and discussed. In the end, AFM is used to analyze the surface roughness of different nanofibers (blend nanofibers, pure PVA and nanofibers after selective dissolution). The results show the entirely removal of PVP from PVA and the successful preparation of surface roughened PVA nanofibers. The AFM testing result demonstrates a huge improvement in surface roughness after the selective dissolution procedure.

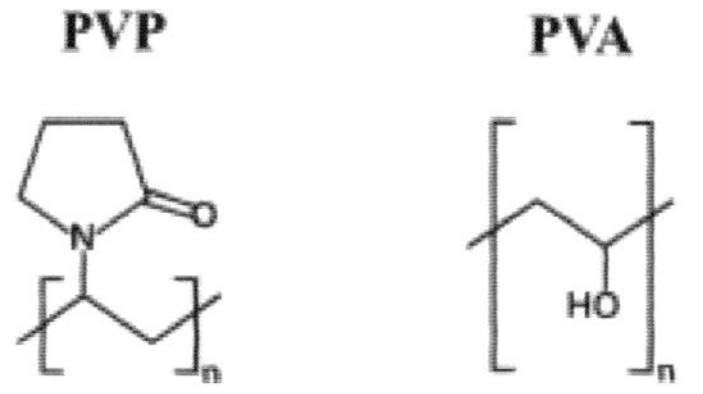

Figure 1. Molecular structures of PVP and PVA.

2. Sample preparation

2.1. Materials

PVA powder from Aladdin Co. Ltd. China, specified as #1799 (M_w = 74885), is selected because it dissolves well in warm water (typically at temperatures above 95°C), but can hardly dissolve in cold water.

PVP powder from Chron Co. Ltd. China, specified as #K30 (M_w = 44000 ~ 54000), is used, since it shows good solubility in cold water and ethanol. Ethanol was purchased from Chuandong Chemical (Chongqing, China). Ultra pure water (resistivity higher than 1.8 GΩ·m at 25°C) was produced by the ultra pure water system in lab.

2.2. Apparatus

Electrospinning machine (Aibo Zhiye Ion Technology Co., Ltd., Beijing, China) was used to generate nanofibers. Nanofiber morphology testing and EDS analysis were operated on scanning electron microscope(SEM)(Quattro S, Thermo Fisher Scientific Inc., USA). Roughness of nanofiber surface was measured by AFM (MFP-3D-BIO, Asylum Research Inc., USA).The diameters of nanofibers were measured according to SEM photos. X-ray photoelectron spectroscopy (XPS) testing is carrued out by the equipment:ESCALAB250Xi, Thermo Fisher Scientifific, Waltham, MA, USA.

2.3. Preparation of PVA-PVP blend nanofibers
Electrospun solutions are prepared by mixing 3 g PVA and 3 g PVP into 34 g ultra-pure water. Then the blend was removed on the heating and stirring system at 95°C and stirring at 500 rad/min for 4 hours. Afterwards the solution goes through a cooling process at room temperature for an hour in order to remove the bubbles inside solution. The as prepared solution is injected into a 10 mL syringe and electrospinning runs for 4 hours on the electrospinning machine with applied voltage of 19 kV, solution flow rate of 0.5 mL/h and nozzle-to-collector height of 24 cm. The as gained nanofibers are collected on an aluminum foil attached on the flat collector. Then the blend nanofibers are sent to vacuum drying at 55°C for 4 hours.

2.4. Preparation of surface roughened nanofibers
When the electrospinning is finished, the nanofiber nonwovens are cut into 2 cm×2 cm pieces and every four pieces is immersed into 500 mL ethanol under room temperature for 2 weeks with a stirring speed of 500 rad/min to selectively dissolve PVP. After the selective dissolution in ethanol, the immersed nanofiber pieces are sent to vacuum drying for 4 hours at 55°C. In order to analyse the morphology on nanofiber surface, SEM, FTIR, EDS and XPS analysis is utilized to characterize nanofiber samples.

3. Testing results

3.1. Transform Infrared Spectra (FTIR) Analysis.
FTIR is an equipment to characterize the composition of samples. Thanks to its convenience and simple operation, it has been widely used to check functional groups of sample by checking characteristic peaks of samples. According to ref[18, 19], there exists a series of peaks such as the peak from 3550 to 3200 cm^{-1}, peaks at 2933 cm^{-1}, 1647cm^{-1}, 1092cm^{-1} in PVA FTIR spectra as Table 1 shows. These peaks correspond to O-H stretching bands from the intermolecular and intramolecular hydrogen bonds, C-H stretching bands, C=O stretching bands and C-O stretching bands respectively. Table 2 exhibits characteristic peaks of PVP. The major peaks of PVP locate at 1660, 1440, 1370, 1290 cm^{-1}, which correspond to C=O stretching bands, O-H stretching bands, lactone structures, C-N stretching bands[20, 21].

Table 1. Assignments of peaks in FTIR spectrum of pure PVA polymer nanofibers.

Sample	Wavenumber(cm^{-1})	Spectral Band
Pure PVA nanofibers	3200 to 3550	O-H Stretching band
	2933	C-H Stretching band
	1647	C=O stretching band
	1092	C-O stretching band

Table 2. Assignments of peaks in FTIR spectrum of pure PVP polymer nanofibers.

Sample	Wavenumber(cm^{-1})	Spectral Band
Pure PVP nanofibers	1660	C=O stretching band
	1440	O-H stretching band
	1370	lactone structure
	1290	C-N stretching band

As Figure 2 shows, the curve of blend nanofibers contains both characteristic peaks of PVA and PVP, which demonstrates the successive preparation of PVA-PVP blend nanofibers. Compared with pure PVA, the peak of PVA-PVP blend around 1652 cm^{-1} shifts to lower wavenumbers because hydrogen bond forms between PVA's hydroxyls and PVP's carbonyls[22]. As for curve of sample after selective dissolution, the disappearance of peak at 1290 cm^{-1} and sharp decrease of peak at 1652 cm^{-1} indicate the complete removal of PVP composition from blend nanofibers. The sample after selective dissolution dose not show any peaks of PVP compared with blend nanofibers sample. Furthermore, after the procedure of ethanol immersing, the peaks of PVA all remain, such as peaks from 3200 to 3550 cm^{-1}, 2933 cm^{-1} and 1092 cm^{-1} which denotes O-H bands, C-H bands and C-O bands in PVA. Result mentioned above means most of PVP has been washed out and the composition of nanofiber now is nearly pure PVA.

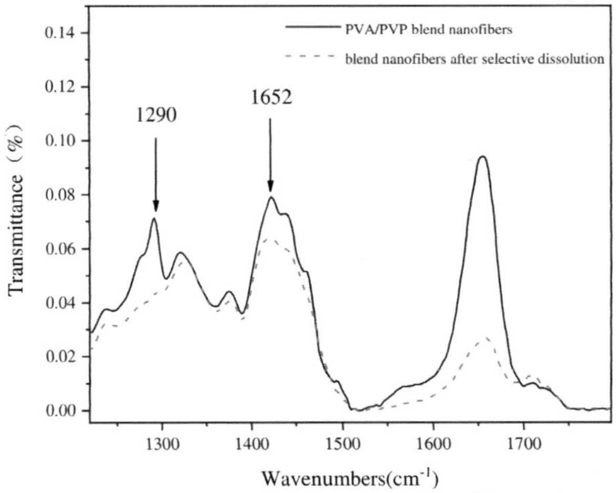

Figure 2. FTIR spectrum of PVA/PVP blend and blend nanofibers after selective dissolution.

3.2. Energy Dispersive Spectrometer (EDS) Chemical Analysis.

It is a good method to measure element changes in order to find out the amount of PVP left in the blend nanofiber after selective dissolution considering PVA only contains carbon, oxygen and hydrogen three elements but PVP includes nitrogen element besides the three elements. Thus C, O and N are set as main chemical elements for EDS to test quantitatively. During the EDS testing, five points are taken randomly from PVA-PVP blend nanofibers and blend nanofibers after selective dissolution. The results of EDS chemical analysis of PVA-PVP blend nanofibers and blend nanofibers after selective dissolution are displayed in Figure 3. The appearing of N element in Figure 3(a) indicates the fluky production of PVA-PVP blend nanofibers. The N element takes about 11.5% in the composition of PVA-PVP blend nanofibers as it revealed in Table 3(a). According to the curve-similarity of five random points on PVA-PVP blend nanofibers in Figure 3(a), a conclusion can be made: PVA and PVP mixed evenly on fiber, which is the precondition to form rough structure on nanofiber surface. However as it illustrates in Figure 3(b) and Table 3(b), when the PVA-PVP blend nanofiber goes through selective dissolution, the N element amount in the sample decreased sharply to 1.9%, which means most of PVP has been successively removed. The weight ratio of C to O in surface roughened nanofibers is in close agreement to the weight ratio of C to O of pure PVA [23].

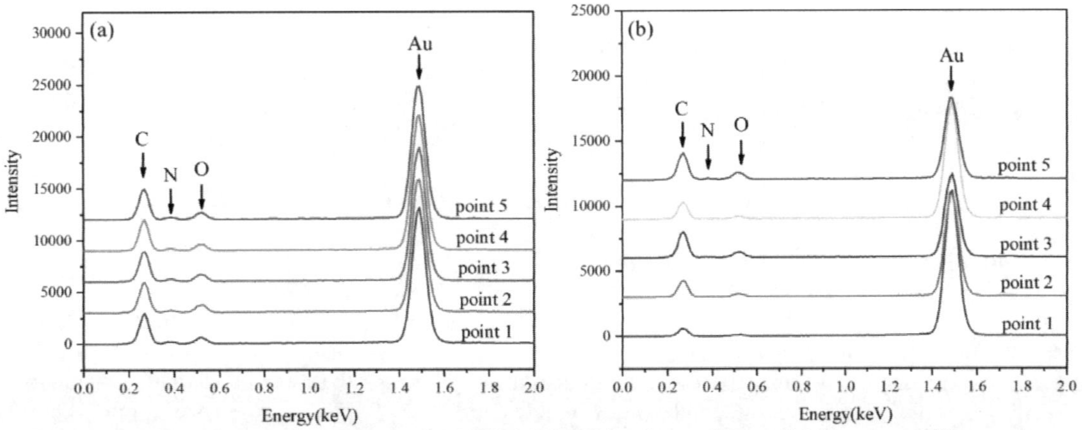

Figure 3. EDX chemical analyses of (a) PVA-PVP blend nanofibers (b) blend nanofibers after selective dissolution.

Table 3. EDX chemical composition of PVA-PVP blend nanofibers and blend nanofibers after selective dissolution:(a) PVA-PVP blend nanofibers,(b) blend nanofibers after selective dissolution

(a)			(b)		
Element	Weight%	Atomic%	Element	Weight%	Atomic%
C	60.52	66.16	C	68.81	74.43
N	12.24	11.47	N	2.03	1.88
O	27.24	22.36	O	29.16	23.68

3.3. Scanning Electron Microscope (SEM) Analysis.

In order to check out the surface roughness change of PVA-PVP blend nanofibers before and after selective dissolution, a high-resolution SEM are adopted to check nanofiber samples. Magnifications of 5000 X is applied during the testing process. To avoid unnecessary damage to nanofibers, the accelerating voltage of SEM electron beam is set to 20 kV. The morphology of blend nanofibers (weight ratio of PVA to PVP=1:1) before and after selective dissolution is displayed in Figure 4.

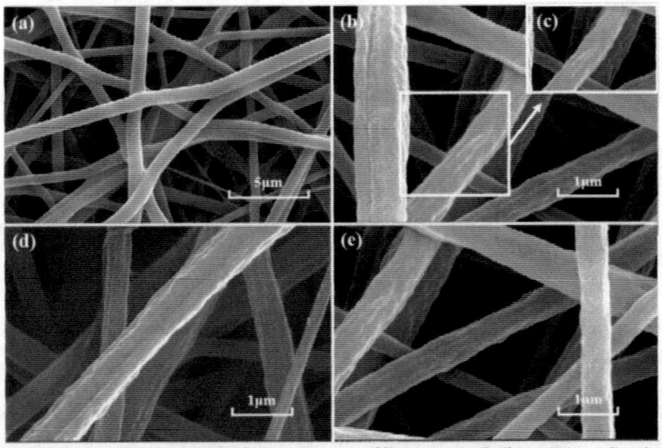

Figure 4. SEM photos of (a) PVA-PVP blend nanofibers and (b), (c), (d), (e) surface roughened nanofibers generated through selective dissolution of PVA-PVP blend nanofibers.

The surface of blend nanofibers before selective dissolution is smooth without any pore or stratification as Figure 4(a) shows, which means PVA and PVP distribute homogeneously in every nanofibers rather than separately. The average diameter of the PVA-PVP blend nanofibers is about 650nm. The homogeneously distribution of PVA and PVP creates condition to remove PVA evenly and leave regular structures on nanofibers surface. Morphology of one single PVA-PVP blend nanofiber after selective dissolution has been illustrated in Figure 4(b), (c). It is obvious the surface of nanofiber after selective dissolution is cucumber-like, rough and full of grooves and depressions due to the disintegration of PVP. The average diameter of the PVA-PVP blend nanofibers after selective dissolution decreases to 600 nm because of the loss of PVP. Though all areas for SEM scanning are randomly selected, all nanofibers within chosen areas are surface-roughened as Figure 4(d) and (e) show. The result directly proves the effectiveness of selectively removal of PVP in creating surface roughened PVA nanofibers, which supports the result of testings mentioned above.

3.4 Atomic Force Microscope (AFM) Analysis.
The rough surface of nanofiber has been proved by SEM testing, however more analysis should be taken to find out the roughness change quantitatively. Atomic Force Microscope (AFM), which is a three-dimensional topographic equipment with a high atomic resolution, is adopted to check roughness of blend nanofibers before and after selective dissolution. Because the average diameter of nanofiber is nearly 1 μm obtained under the condition of weight ratio PVA:PVP=1:1, electrospinning voltage 19 kV, flow rate 0.5 mL/min, the AFM scan range is set as 1μm×1μm. And the roughness is indicated by using the difference between Max height and Min height, which is defined as H-value. According to the results of AFM in Figure 5(a) and (b), the H-values of PVA-PVP blend nanofibers before and after selective dissolution are 95.7 nm and 652.2 nm respectively. The roughness of nanofibers after selective dissolution is about 5.8 times higher than PVA-PVP blend nanofibers before selective dissolution. Judging from The 3D morphology graphs of PVA-PVP blend nanofiber before and after selective dissolution, it is obvious the nanofibers after selective dissolution show more ups and downs compared with the ones before selective dissolution. The testing result of AFM demonstrates the significant increase of nanofiber roughness after selective dissolution. AFM result demonstrates that selective removal of PVP from PVA-PVP blend nanofibers makes surface of PVA-PVP blend nanofibers more rough.

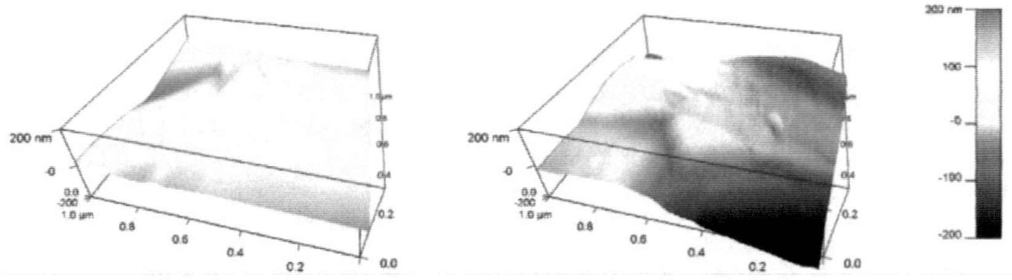

(a) (b)

Figure 5. AFM topographic maps of PVA/PVP blend nanofibers (a) before selective dissolution and (b) after selective dissolution.

4. Discussion
It is well known the molecule transportation influences the component distribution of blend system. Enough energy is a precondition for molecule transportation. So the surface component distribution under different vacuum drying temperature is researched. XPS, a highly sensitive surface analysis method with typical probing depth of around 5~10 nm, is therefore pretty suitable for the surface compositional analysis of nanofibers. As shown in Figure 6, The peaks at 285, 399, and 531 eV are representations of C1s, N1s, and O1s, respectively. By evaluation on the peak areas of C1s, O1s and N1s in Figure 6, the atomic ratio of N for blend nanofiber samples before and after vacuum drying is 3.14% and 3.66% respectively, which demonstrated more component-even nanofiber surfaces are formed after the vacuum drying. The principle behind is about the state of minimum energy dissipation and mobility of PVP. During the electrospinning, the electrostatic repulsion caused by surface charges prolongs the PVA/PVP blend droplet into a Taylor cone, from which a charged jet is ejected. The jet initially extends in a straight line and then undergoes whipping motions. Upon the

whipping process, most water inside fibers evaporates quickly. So the PVA tends to crystallize from water because of its low solubility and forms the outer surface of blend nanofiber. At the same time, PVP and some water still exist as solution inside the fibers. The subsequent vacuum drying treatment endowed polymer molecules enough energy and the two polymer both have increased mobility. The lower viscosity fluid tends to migrate to outer surface of the blend nanofibers system[24].So the PVP, being characterized by high mobility because of lower Mw[25, 26] , tends to segregate to the surface of the fibers because of the lower surface tension to reach a minimum energy state of the blend system. At the same time, PVA remain stable at initial distribution because of weak mobility due to its high Mw. Though the PVP component generates a directional migration to fiber surface driven by decreasing the system free energy, the massive hydrogen band between two polymers impedes the extensive PVP migration[27]. So the surafce of the blend nanofiber is formed by both PVA and PVP. By remove PVP selectively, surface-roughened PVA nanofiber is achieved.

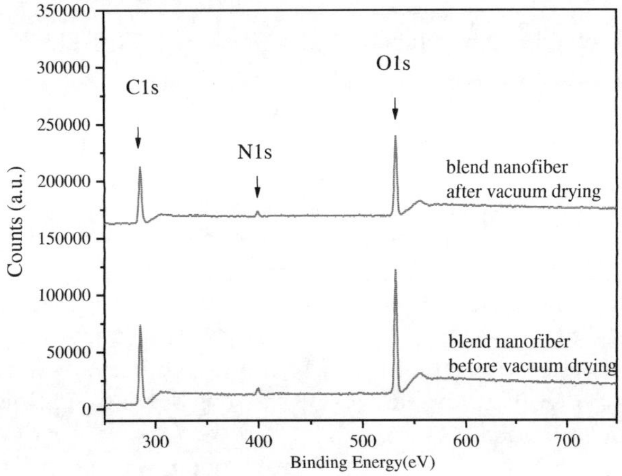

Figure 6. XPS spectra of nanofiber samples before and after selective dissolution.

5. Summary

To conclude, PVA-PVP blend nanofibers have been successively prepared. The nanofibers show a nanoscale diameters ranging from 400 nm to 800 nm. Unlike other polymer combinations, lower Mw polymer (PVP) has been removed by selective dissolution to achieve rough surface. Ethanol has been chosen as solvent to remove PVP by utilizing the solubility difference of PVA and PVP in it. This is also the first time to create surface roughened nanofibers by using PVA and PVP combination. In order to make sure all PVP composition is washed out from the blend, testing like FTIR and EDS has been carried out. Obvious changes of FTIR characteristic peaks position and intensity, the sharply decreased N element after selective dissolution demonstrate the complete removal of PVP from blend nanofibers successively. SEM and AFM are adopted to understand the surface morphology and surface roughness, pores etc. Testing results in this work confirm the effectiveness of selective dissolution to create surface roughened PVA nanofibers. By XPS analysis, the vacuum drying process is foud to be an important factor for the formation of surface-roughened nanofibers. The mobility difference of PVA and PVP molecule causes separated layers on nanofiber surface during solidification process. The surface-roughened nanofibers could be applied in water filters, cell host matrixes, etc.

6. References

[1] Din I U, Shaharun M S, Naeem A, Alotaibi M A, Alharthi A I, Bakht M A and Nasir Q 2020 Carbon nanofibers as potential materials for catalysts support, a mini-review on recent advances and future perspective *Ceram. Int.* **46** 18446-52

[2] Mane P P, Ambekar R S and Kandasubramanian B 2020 Electrospun nanofiber-based cancer sensors: A review *Int. J. Pharm.* **583** 119364

[3] Kalantari K, Afifi A M, Jahangirian H and Webster T J 2019 Biomedical applications of chitosan electrospun nanofibers as a green polymer–Review *Carbohydr. Polym.* **207** 588-600
[4] Xue J, Wu T, Dai Y and Xia Y 2019 Electrospinning and electrospun nanofibers: Methods, materials, and applications *Chem. Rev.* **119** 5298-415
[5] Al-Enizi A M, Zagho M M and Elzatahry A A 2018 Polymer-based electrospun nanofibers for biomedical applications *Nanomaterials* **8** 259
[6] Monfared M, Taghizadeh S, Zare-Hoseinabadi A, Mousavi S M, Hashemi S A, Ranjbar S and Amani A M 2019 Emerging frontiers in drug release control by core–shell nanofibers: A review *Drug Metab. Rev.* **51** 589-611
[7] Huang W, Tong Z, Wang R, Liao Z, Bi Y, Chen Y, Ma M, Lyu P and Ma Y 2020 A review on electrospinning nanofibers in the field of microwave absorption *Ceram. Int.* **46** 26441-53
[8] Jalaja K, Naskar D, Kundu S C and James N R 2016 Potential of electrospun core–shell structured gelatin–chitosan nanofibers for biomedical applications *Carbohydr. Polym.* **136** 1098-107
[9] Homaeigohar S, Davoudpour Y, Habibi Y and Elbahri M 2017 The electrospun ceramic hollow nanofibers *Nanomaterials* **7** 383
[10] Yang Y, Chang S, Bai Y, Du Y and Yu D-G 2020 Electrospun triaxial nanofibers with middle blank cellulose acetate layers for accurate dual-stage drug release *Carbohydr. Polym.* **243** 116477
[11] Li Z, Xu Y, Fan L, Kang W and Cheng B 2016 Fabrication of polyvinylidene fluoride tree-like nanofiber via one-step electrospinning *Materials & design* **92** 95-101
[12] Philip P, Jose E T, Chacko J K, Philip K and Thomas P 2019 Preparation and characterisation of surface roughened PMMA electrospun nanofibers from PEO-PMMA polymer blend nanofibers *Polym. Test.* **74** 257-65
[13] Kim G-M, Le K H T, Giannitelli S M, Lee Y J, Rainer A and Trombetta M 2013 Electrospinning of PCL/PVP blends for tissue engineering scaffolds *Journal of Materials Science: Materials in Medicine* **24** 1425-42
[14] Kim M N, Koh J, Lee Y and Kim H 2009 Preparation of pva/pan Bicomponent nanofiber via electrospinning and selective dissolution *J. Appl. Polym. Sci.* **113** 274-82
[15] You Y, Youk J H, Lee S W, Min B-M, Lee S J and Park W H 2006 Preparation of porous ultrafine PGA fibers via selective dissolution of electrospun PGA/PLA blend fibers *Materials Letters* **60** 757-60
[16] Moon S, Choi J and Farris R J 2008 Highly porous polyacrylonitrile/polystyrene nanofibers by electrospinning *Fibers and Polymers* **9** 276
[17] Mercader C, Lucas A, Derré A, Zakri C, Moisan S, Maugey M and Poulin P 2010 Kinetics of fiber solidification *Proceedings of the National Academy of Sciences* **107** 18331-5
[18] Abdelrazek E, Elashmawi I and Labeeb S 2010 Chitosan filler effects on the experimental characterization, spectroscopic investigation and thermal studies of PVA/PVP blend films *Physica B: Condensed Matter* **405** 2021-7
[19] Mansur H S, Sadahira C M, Souza A N and Mansur A A 2008 FTIR spectroscopy characterization of poly (vinyl alcohol) hydrogel with different hydrolysis degree and chemically crosslinked with glutaraldehyde *Materials Science and Engineering: C* **28** 539-48
[20] Baganizi D R, Nyairo E, Duncan S A, Singh S R and Dennis V A 2017 Interleukin-10 conjugation to carboxylated PVP-coated silver nanoparticles for improved stability and therapeutic efficacy *Nanomaterials* **7** 165
[21] Deniz A E, Vural H A, Ortaç B and Uyar T 2011 Gold nanoparticle/polymer nanofibrous composites by laser ablation and electrospinning *Materials Letters* **65** 2941-3
[22] Wei Q, Zhang Y, Wang Y, Chai W and Yang M 2016 Measurement and modeling of the effect of composition ratios on the properties of poly (vinyl alcohol)/poly (vinyl pyrrolidone) membranes *Materials & Design* **103** 249-58
[23] Andrade G I, Barbosa-Stancioli E F, Mansur A A P, Vasconcelos W and Mansur H S 2008 Small-angle X-ray scattering and FTIR characterization of nanostructured poly (vinyl alcohol)/silicate hybrids for immunoassay applications *JMatS* **43** 450-63

[24] Wei M, Kang B, Sung C and Mead J 2006 Core - sheath structure in electrospun nanofibers from polymer blends *Macromolecular Materials and Engineering* **291** 1307-14
[25] Mirtič J, Balažic H, Zupančič Š and Kristl J 2019 Effect of solution composition variables on electrospun alginate nanofibers: response surface analysis *Polymers* **11** 692
[26] Vigani B, Rossi S, Milanesi G, Bonferoni M C, Sandri G, Bruni G and Ferrari F 2018 Electrospun alginate fibers: Mixing of two different poly (ethylene oxide) grades to improve fiber functional properties *Nanomaterials* **8** 971
[27] Zhao Q, Qi H J and Xie T 2015 Recent progress in shape memory polymer: New behavior, enabling materials, and mechanistic understanding *Prog. Polym. Sci.* **49** 79-120

Acknowledgments

The authors gratefully acknowledge the equipment support (SEM, XPS equipment, etc.) from Analytical and Testing Center of Chongqing University.

Asphalt Interface Based on Abaqus Asphalt Pavement Temperature Field and Aggregate Morphological Characteristics Measurement System

Zhijun Ying[1*], Huimin Wang[2], Yunfei Zhao[2], Zhihao Wu[3], Sili Shen[1]

[1] Hangzhou Urban Expressway Co., Ltd., Hangzhou, China
[2] Highway College of Chang'an University, Xi'an, China
[3] Zhejiang Expressway Construction Management Co., Ltd., Zhejiang, China

*Corresponding author's e-mail: yingzjok@yeah.net

Abstract. To reveal the adhesion behavior of the aggregate - asphalt interface under water-temperature coupling, the temperature field of the asphalt pavement was simulated and analyzed based on Abaqus, and the morphology of the aggregate was analyzed using the Aggregate Morphological Characteristic Measurement System (AIMS II) to investigate the adhesion behavior of the asphalt interface under water-temperature coupling. The Abaqus software was used to study the temperature distribution pattern of the pavement structure of the dependent project and to construct a regional characteristic distribution model for Zhejiang Province. On this basis, aggregate texture shaping equipment was developed, aggregate morphological characteristics were tested with the aid of the Aggregate Morphological Characteristics Inspection System (AIMS II), the water immersion weighing method was used to evaluate the aggregate - asphalt interface adhesion, and the relationship between aggregate morphological characteristics and macroscopic adhesion was established. The study shows that there is a linear negative correlation between sphericity and mass loss rate, a linear positive correlation between gradient angles and mass loss rate, and an optimum value for aggregate surface texture. Using a modified 'sandwich' pulling test, four lithologies of limestone, basalt, granite and sandstone were selected to investigate the effect of environmental factors such as water and temperature on the adhesion of the aggregate to the asphalt interface. The results showed that the interfacial bonding performance of the aggregate and asphalt increased as the heating temperature of the aggregate increased, while the interfacial bonding performance of the aggregate and asphalt weakened as the water immersion time increased. The research results have certain theoretical guidance significance for enhancing the water damage resistance of asphalt mixes.
Keywords: Abaqus; adhesion water; temperature coupling; Aggregate Morphological Characteristics Inspection System

1. Introduction

As a non-uniform three-phase material, the morphological composition of the asphalt mixture mainly includes aggregate, asphalt, and aggregate-asphalt contact surface. Among them, the aggregate mainly plays the role of bearing the road load, the asphalt mainly plays the role of cementation and filling, the aggregate is cemented into a whole, and the interface as the link between the aggregate and the asphalt interaction and transmission, between the aggregate and the asphalt, the strength is relatively low, easy to be affected by external factors and destroyed [1-3].

The adhesion behavior of the aggregate-asphalt interface is critical to the water stability of asphalt pavement. In high temperature and rainy areas, water will penetrate the aggregate and asphalt interface, displacing the asphalt on the surface of the aggregate, while the temperature will aggravate molecular movement, further promote the damage of the aggregate-asphalt interface, and the water-temperature coupling effect will significantly affect the adhesion behavior between aggregate and asphalt. Therefore, the effect of water-temperature coupling on aggregate-asphalt interface adhesion is of great significance for improving the water stability of asphalt mixtures in high temperature and rainy areas [4-5].

At present, researchers at home and abroad have conducted a large number of studies on the adhesion of the aggregate-asphalt interface. Arabi et al. Used limestone, granite, and quartzite to coat the aggregate by adding two types of polyethylene, high-density, and low-density, and used relevant indoor experiments to evaluate the water damage of the asphalt mixture [6]. Luo Rong et al. used the surface energy theory-related test methods to test the surface energy parameters of different matrix asphalt and aggregates, established the interface failure model under drying conditions and water conditions, and used the water immersion method for verification [7]. Based on indoor experiments, Moraes et al. studied whether the macroscopic pull strength is consistent with the interfacial strength obtained by the surface energy test [8]. Dou Yingying et al. studied the adhesion attenuation mechanism of the asphalt-aggregate interface underwater erosion from a macroscopic and microscopic perspective of surface free energy theory, attenuated total reflection infrared spectroscopy (FTIR-ATR), and atomic force microscopy (AFM) [9]. Sun Jishu et al. Used 70# matrix asphalt and nano-bentonite modified asphalt as test samples, determined the contact angle between asphalt and aggregate under different aging conditions by lying drop method, and calculated the cohesion function of asphalt based on microscopic surface energy theory [10]. It can be seen that for the interface adhesion behavior of aggregate-bitumen, the current research has made some progress in theoretical evaluation, numerical simulation, and mechanism analysis, while quantitative research under the coupling of environmental factors has rarely been reported [11-12].

Based on this, an improved "sandwich" drawing test is used to quantitatively and intuitively study the effect of water and temperature on the adhesion of the aggregate-asphalt interface based on the mechanical response parameters, to comprehensively and comprehensively evaluate the adhesion of the aggregate-asphalt interface and improve the water damage resistance of the asphalt mixture.

Abaqus method for determining pavement temperature. The pavement structure is exposed to the air, and the temperature and solar radiation will affect the internal temperature distribution of the pavement structure. Therefore, the Abaqus finite element software was used to analyze the temperature variation law of the pavement structure of the Hang Shao section of the Western Duplication Project according to the theory related to the temperature field.

Calculation model. Abaqus finite element software was used to analyze the temperature distribution law of the asphalt pavement structure of the relying project. The grid was divided from the middle to both sides from dense to sparse according to a certain ratio, and the calculation model is shown in Figure 1.

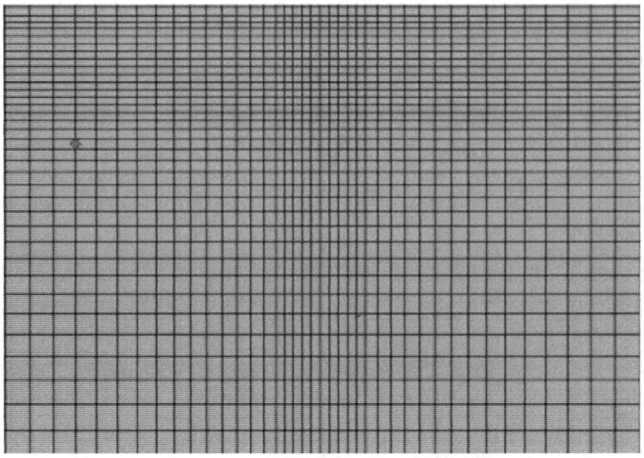

Figure 1. Calculation model

Material parameters. Concerning relevant studies, the material parameters are taken as shown in Table 1.

Table 1. Material parameters

parameters	SMA-13, Sup-20, Sup25	Cement stabilized aggregates	Soil base
Thermal conductivity kJ/(m*h°C)	4680	5616	5616
Density ρkg/m3	2300	2200	1800
Heat capacity C (J/kg*°C)	924.9	911.7	1010.0
Solar emissivity αs		0.90	
Road surface emissivity ε		0.81	
Absolute zero T/°C		-273	

Temperature. Representative temperature data for 1 day and 24h in the hot season in Zhejiang Province were selected, with a minimum of 21.1° C and a maximum of 35.9° C.

Solar radiation. In Abaqus, the effect of radiation on the pavement temperature field is mainly achieved using the subroutine DFLUX.

Convective heat transfer. The survey found that the wind speed is generally high in Zhejiang Province during the high-temperature season, with a wind speed v of 3.7m/s. The convective heat exchange between the pavement and the atmosphere is realized in Abaqus using the subroutine FILM.

Ground temperature. According to the survey, it was found that the daily minimum temperature in Zhejiang Province was 22.6° C and the daily maximum temperature was 35.9° C.

2. The Test Process of the Effect of Aggregate Morphological Characterization System Used in the Asphalt Interface

2.1. Aggregate morphological characterization system (AIMS II) to collect aggregate morphology

The components of the Aggregate Morphology Measurement System (AIMS II) consist of a set of hardware and a computer, where the hardware platform is used to collect aggregate images and the computer is used to run the system and carry out data analysis. The hardware platform for aggregate image acquisition consists of a camera, an optical microscope, an aggregate tray, a backlighting system, and a top lighting system. The computer is equipped with the AIMS software for system operation management and analysis and output of data results. The entire measurement system is shown in Figure 2.

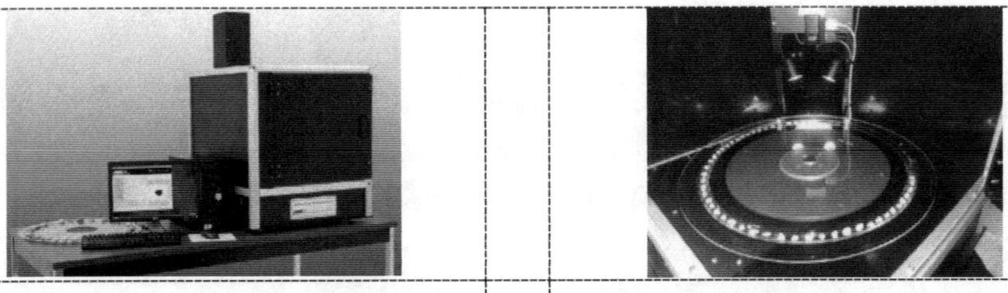

(a) External construction drawings | (b) Internal construction drawing

Figure 2. Aggregate Image Measurement System (AIMS II)

2.2. AIMS II evaluation indicators

(1) Sphericity

The sphericity index is used in AIMS II to evaluate the three-dimensional shape of the coarse aggregate. The sphericity index is calculated from the three-dimensional dimensions of the aggregate using equation (1), as shown in Figure 3.

$$S_p = \sqrt[3]{\frac{d_S d_I}{d_L^2}} \quad (1)$$

where: d_S - Aggregate short axis length (height), mm; d_I -aggregate secondary axis length (width), mm; d_L -aggregate long axis length (length), mm.

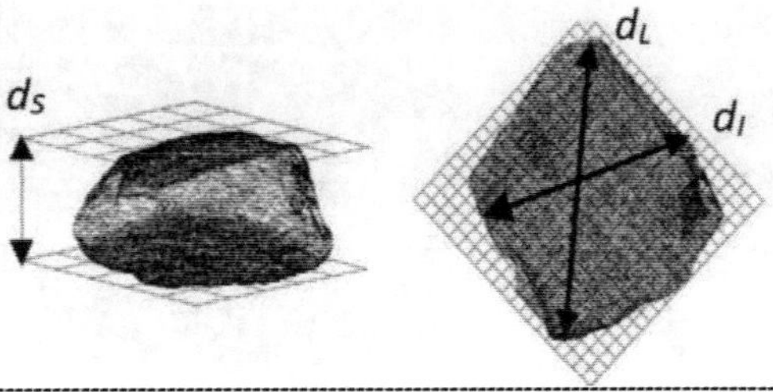

Figure 3. Three-dimensional dimensions of the aggregates.

(2) Gradient Angularity

The angularity of the aggregates in AIMS II is quantified using the gradient angularity index as shown in Figure 4 and calculated using the equation shown in equation (2).

$$GA = \frac{1}{\frac{n}{3}-1}\sum_{i=1}^{n-3}\left|\theta_i - \theta_{i+3}\right| \quad (2)$$

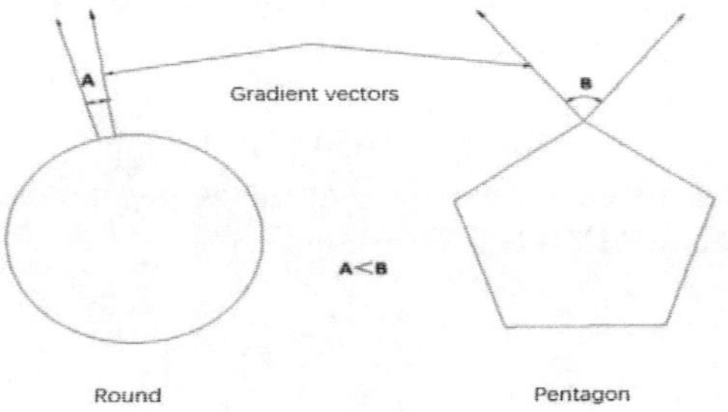

Figure 4. Smooth aggregate versus angular aggregate gradient vectors.

where: θ - Gradient vector angle of the edge points of an aggregate 2-D image n -the total number of points on the edge of the aggregate image; i -the first point of the edge of the aggregate image.

The gradient angularity indicator has a maximum value of 1000 and a minimum value of 0. The angularity classification levels are shown in Figure 5.

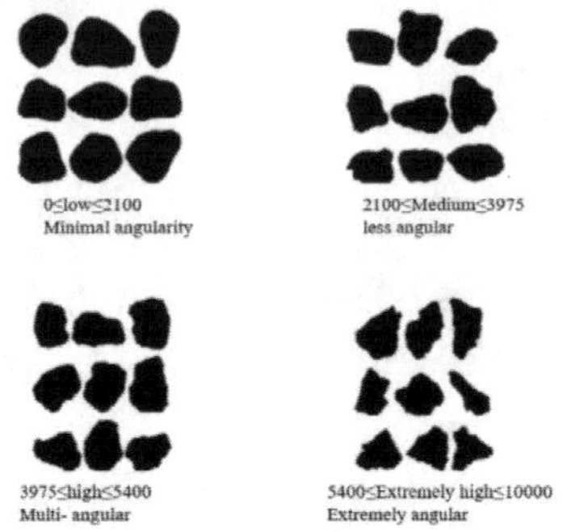

Figure 5. Aggregate angularity classification.

(3) Texture

The surface texture of the aggregate is obtained and quantified by wavelet analysis in AIMS II. The principle of wavelet analysis is shown in Figure 6. and the formula for calculating the surface texture of the aggregate is shown in equation (3).

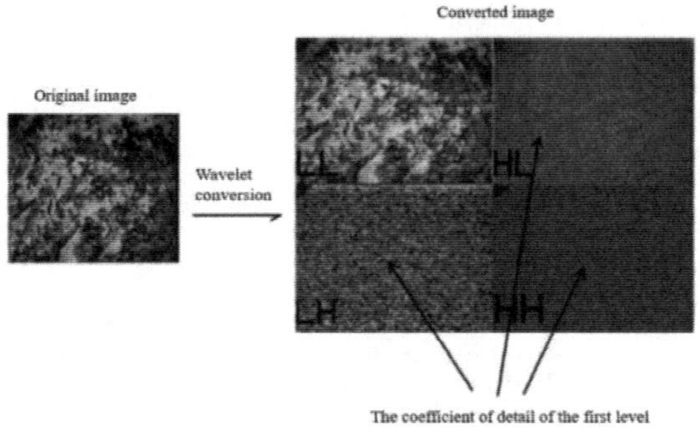

Figure 6. Wavelet analysis.

$$TX = \frac{1}{3N}\sum_{i=1}^{3}\sum_{j=1}^{N}(D_{i,j}(x, y))^2 \tag{3}$$

where: D - the decomposition function; N -the total number of detail coefficients in an image; i - the first high-precision image; j -the wavelet index; x, y -the horizontal and vertical coordinates of the detail coefficients in the transform domain.

The maximum value of the texture index is 1000 and the minimum value is 0. The smaller the value of the texture index, the smoother the surface of the aggregate. the specific classification of the surface texture of the aggregate in AIMS II is shown in Figure 7.

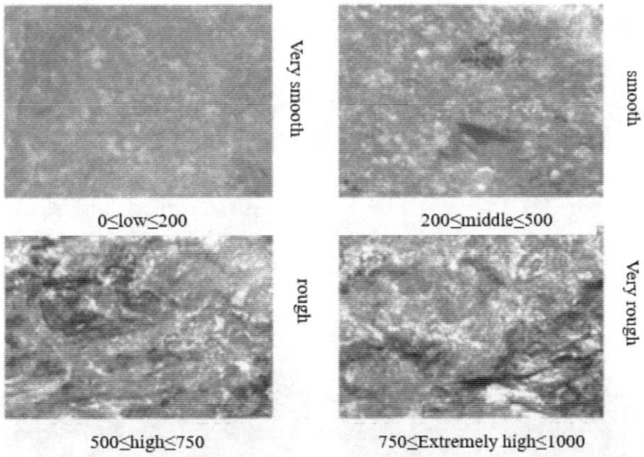

Figure 7. Classification of aggregate surface textures.

2.3. Influence of aggregate shape on the adhesion of aggregate - asphalt interface

To investigate the effect of aggregate shape on the adhesion of aggregate - asphalt interface, by artificially selecting different shapes of aggregates such as cubic, needle, and flake, aggregates of 9.5-13.2mm from 4 different lithologies were divided into 5 groups, and the sphericity index values of different shapes of aggregates were tested by AIMS Ⅱ, to obtain aggregates with different sphericity ranges, according to the classification rules of morphological characteristics index of AIMS. 7 in each group, using limestone as an example, as shown in Figure 8.

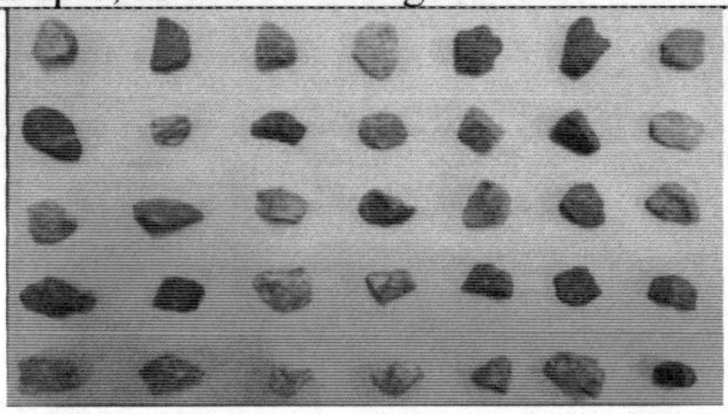

Figure 8. Classification of typical aggregate shapes in limestone.

2.4. Adhesion evaluation indicators

Through the test, the destructive load conditions of different rock aggregates and different conservation conditions can be obtained. To better and more intuitive surface signs of the change of the damage load, and reflect its resistance to water damage performance [13], this paper defines the load loss rate, to obtain the damage of four aggregates under different water bath time conditions, the damage load loss rate is expressed as a formula (4):

$$\alpha_i = \frac{F - F_i}{F} \times 100\% \tag{4}$$

Wherein: for the failure load loss rate; for the dry condition of the failure load; for the water immersion condition of the failure load. The entry of moisture into the asphalt-aggregate interface will cause the adhesion performance of the asphalt mixture to decrease, resulting in early diseases, generally speaking, there are two forms of destruction, adhesion failure, and adhesion failure, and the failure load directly reflects the adhesion performance between the aggregate and the asphalt from a mechanical point of view [14]. On this basis, to analyze the destructive behavior between aggregate and asphalt more intuitively, the peeling rate index (the ratio of the area of asphalt peeling of the surface of the aggregate after the rupture to the total area of the aggregate) is used to evaluate the adhesion performance of asphalt and aggregate [15].

After the pulling fracture, the aggregate surface is shown in Figure 9, the surface of the aggregate is photographed by using a high-definition camera, and then the image processing is carried out through image Pro-Plus software, the number of pixels on the exposed part of the aggregate surface and the number of pixels on the entire aggregate surface are calculated, and the asphalt peeling rate of the asphalt-aggregate interface is calculated according to Formula (5).

$$\beta = \frac{\text{The area of the bare aggregate part}}{\text{The entire interface area}} \qquad (5)$$

Figure 9. Drawing interface.

2.5. Technical performance of raw materials

Asphalt. In this article, Shell 70# matrix bitumen is selected, and its performance indicators are shown in Table 2 [11-12].

Table 2. Technical index of 70# matrix asphalt

Test Indicator	Unit	Standard Range	Test Results	Test Method
Penetration(100g, 25°C, 5s)	0.1mm	60-80	68	T0604
Softening Point TR&B(Universal Law)	°C	≥46	47.4	T0606
Ductility(10°C, 5cm/min)	cm	≥20	33	T0605
Ductility(15°C, 5cm/min)	cm	≥100	115	T0605

From Table 1, it can be seen that the selected asphalt indicators meet the requirements of the specification.

3. Results and discussion

3.1. Pavement structure temperature and aggregate morphology characteristics

Figure 10 is a cloud plot of the temperature distribution of the internal structure of the asphalt pavement, which shows the variation of the temperature of the internal structure of the asphalt pavement with time and depth.

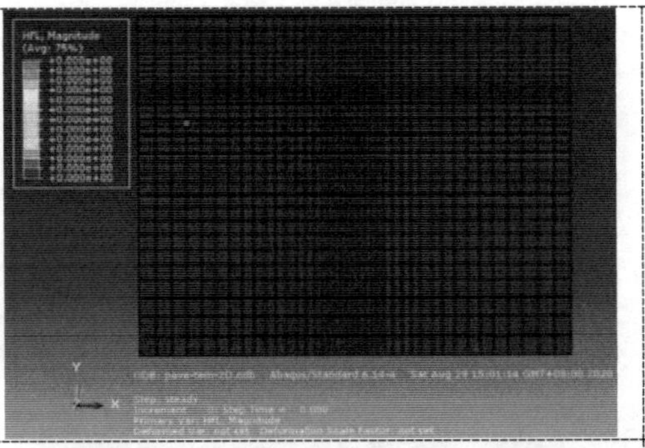

(a) 0h

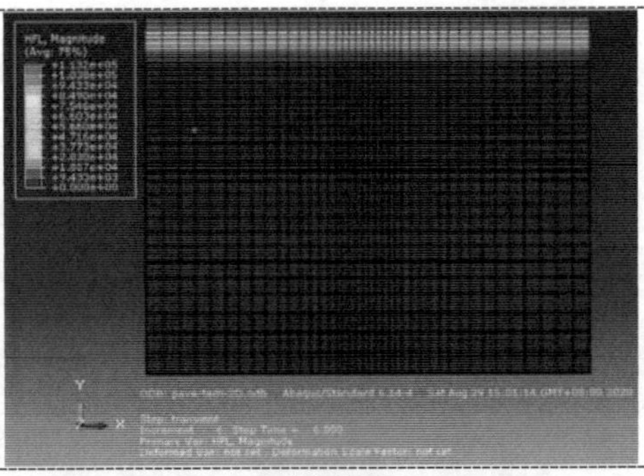

(b) 6h

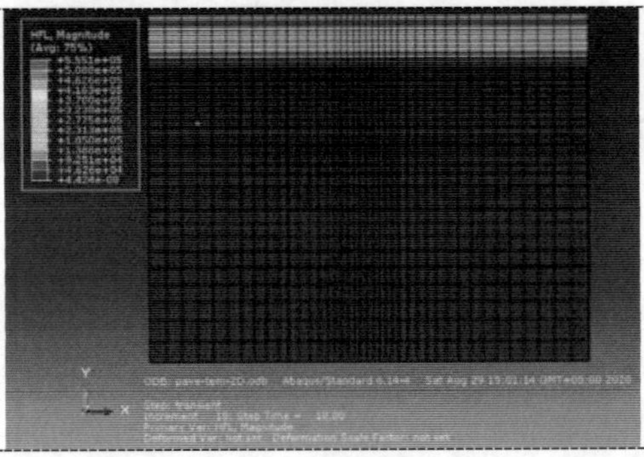

(c) 12h

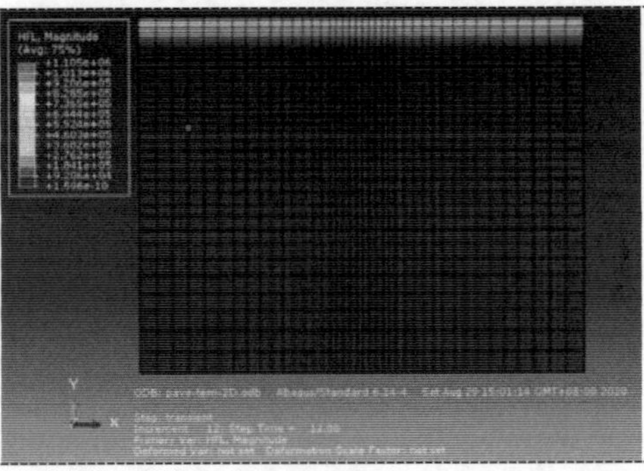

(d) 18h

Figure 10. Temperature distribution of asphalt pavement

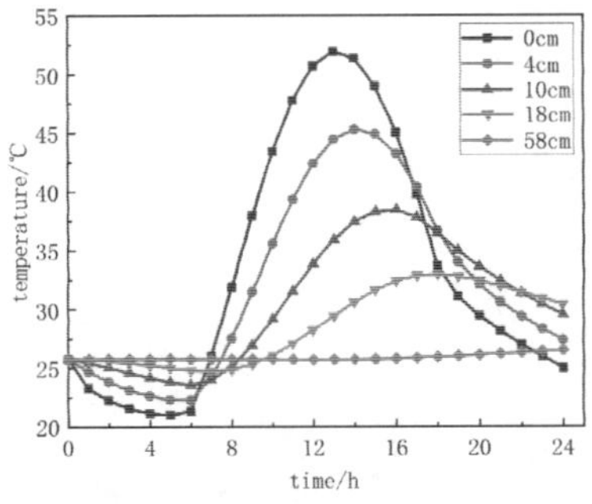

(a) Temperature distribution at different moments

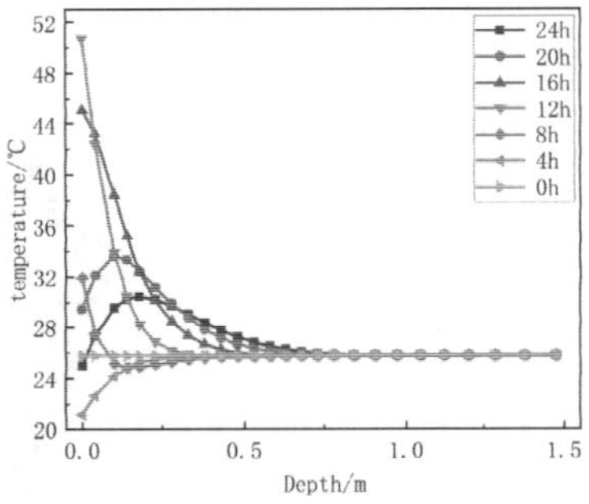

(b) Temperature distribution at different depths

Figure 11. Temperature distribution of asphalt pavement

As can be seen from Figure 11, the internal temperature of the asphalt pavement structure is significantly influenced by time and depth. At 4 am, the temperature at different depths inside the pavement structure was the same; at around noon, due to the increasing atmospheric temperature and solar radiation, the temperature at different depths inside the pavement structure rose sharply to reach its peak, and the rise in road surface temperature was much greater than that at other depths, with the increase in the depth of the pavement structure, the time for different layers to reach the peak temperature gradually lagged; the temperature at different depths inside the pavement structure The temperature at different depths of the pavement structure also varies, with the increase in the depth of the pavement structure, the temperature of each structural layer gradually decreases, for the layer below 50cm of the pavement structure the temperature does not change.

3.2. Influence of aggregate shape on the adhesion of the aggregate - asphalt interface

The results of the AIMS tests are shown in Figure 12, where the first group of aggregates has small sphericity and the fifth group has the largest angularity. The gradual increase in sphericity from group one to group five allows the effect of shape on the aggregate-interface bonding properties to be investigated separately. It is evident from Figure 12 that the angularity and texture values vary slightly for aggregates of the same origin, with significant differences in the angularity and texture inventories for different sources of aggregates, with granite having the largest angularity and texture values and limestone having the smallest angularity and texture values.

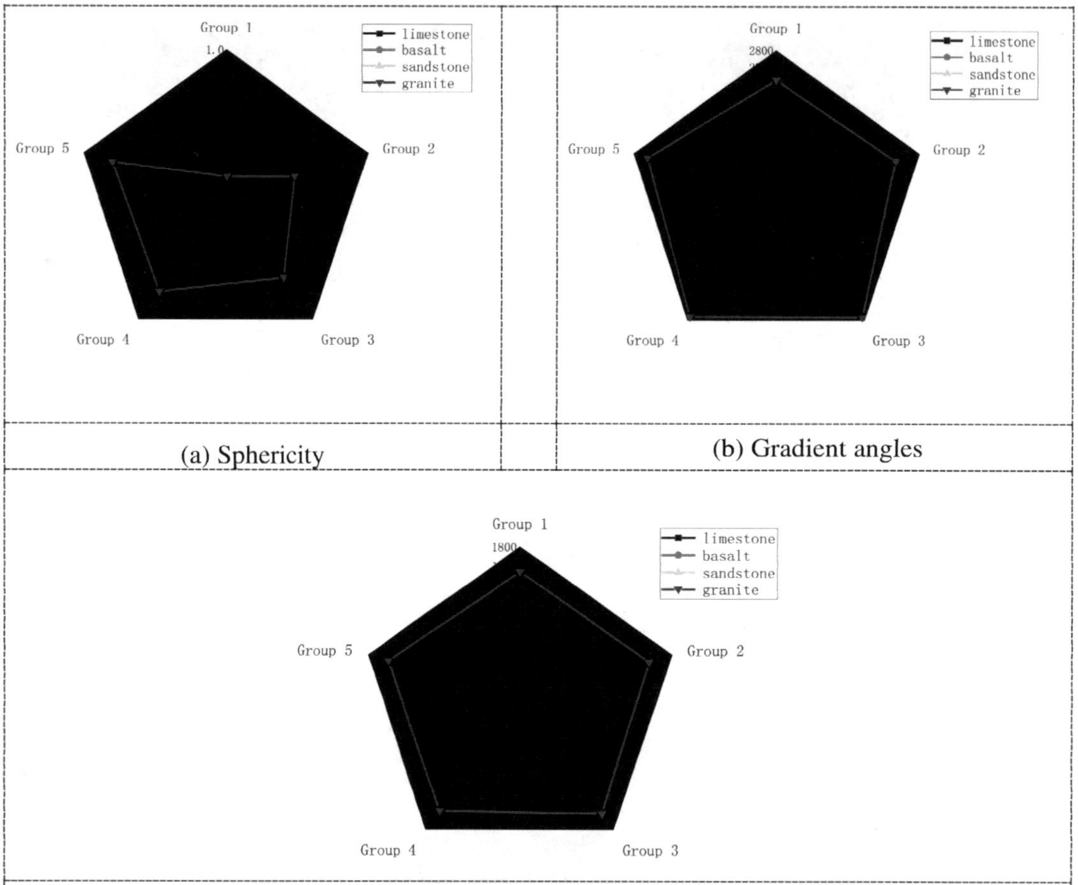

(a) Sphericity

(b) Gradient angles

(c) Surface texture

Figure 12. Morphological characteristics of coarse aggregates.

To investigate the effect of aggregate angularity on the adhesion of the aggregate - asphalt interface, aggregates with five gradient angles of four lithologies were subjected to the water immersion weighing method to establish the fitting relationship between gradient angularity and mass-loss rate, as shown in Figure 12. As can be seen from Figure 13, the larger the gradient angles, the greater the mass-loss rate, indicating that the richer the aggregate angles, the less likely asphalt will adhere to the aggregate surface, and that overly sharp aggregate angles have a detrimental effect on aggregate - asphalt interfacial bonding.

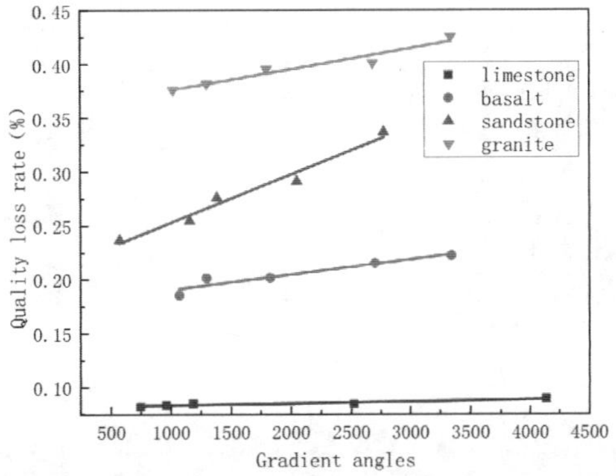

Figure 13. Gradient angles and mass-loss rates.

3.3. Effect of aggregate heating temperature on asphalt-aggregate adhesion performance

The heating temperature of the aggregate affects the degree of expansion of the pores on the surface of the aggregate, affects the degree to which the asphalt penetrates the interior of the aggregate, and further affects the bonding effect between the aggregate and the asphalt [16]. Therefore, the aggregates of limestone, basalt, sandstone, and granite are placed in an oven at 80, 100, 120, 140, and 160 ° C for 1 h, and then the "sandwich" specimen is made according to the steps in the above 1.1, and then the "sandwich" specimen is placed in a 40 ° C tank with water immersion 0, 6, 12, 24h respectively after the immersion is completed, the specimen is taken out and placed in the 15 ° C material testing machine UTM-30 insulation for 30min and then begins the drawing test. The failure load and the failure load loss rate of the specimen are recorded, and the bitumen peeling rate on the aggregate surface is calculated by taking pictures, and the results are shown in Figure 14.

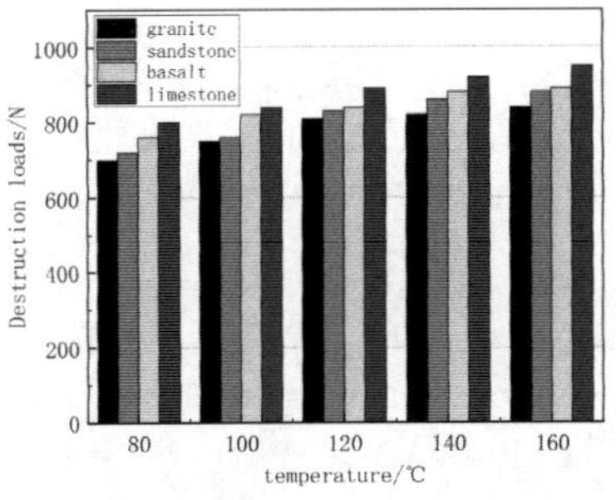

(a) No water conditions

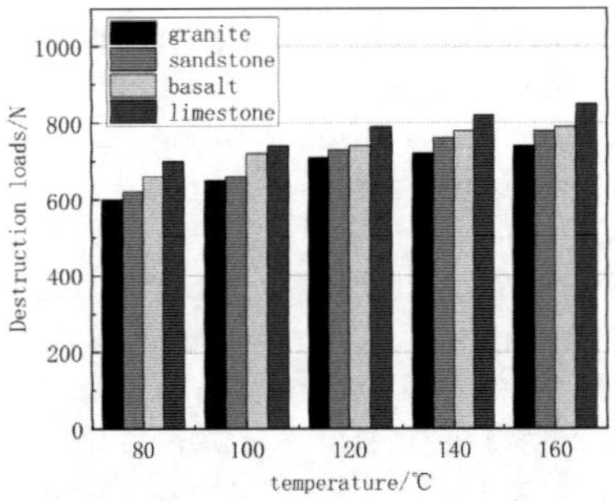

(b) Soak in water for 6h

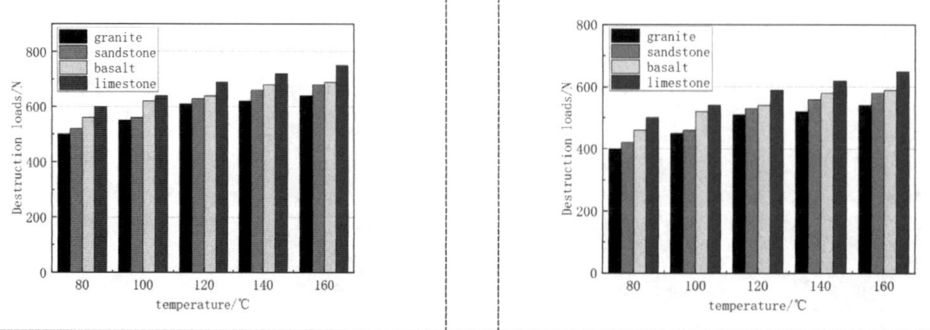

(c) Soak in water for 12h | (d) Soak in water for 24h

Figure 14. The failure load of the interface between aggregate and asphalt under different soaking times and heating temperature.

As can be seen from Fig.14 (a), under dry conditions, the destructive load between the asphalt and the aggregate gradually increases with the increase of the heating temperature of the aggregate. When the heating temperature is increased from 80 ° C to 160 ° C, the aggregate failure load of the four lithic properties increases by an average of about 18%, mainly because of the cohesion failure between the aggregate and the asphalt under dry conditions, and the size of the failure load is mainly related to the adhesion of the asphalt. Fig. 14 (b) ~ Fig. 14 (d) It can be seen that with the prolongation of the immersion time, the adhesion failure occurs at the interface of the aggregate and the asphalt, and the destructive load between the aggregate and the asphalt is gradually reduced, indicating that the presence of water weakens the bonding between the aggregate and the asphalt. The limestone aggregate and asphalt have the largest failure load and the least granite, which is mainly due to the strongest alkalinity of limestone, the weak acidity of asphalt, and the adhesion between limestone and asphalt is stronger than that of granite. At the same time, with the increase of the heating temperature of the aggregate, the destructive load of adhesion failure between the aggregate and the asphalt gradually increases, which is mainly because the pores on the surface of the aggregate are heated and expanded after the aggregate is heated, which is conducive to the penetration of asphalt into the aggregate, and as the temperature returns to normal temperature, the aggregate pores will shrink, resulting in the formation of negative pressure inside the pores on the surface of the aggregate. In addition, the heating temperature of the aggregate is high, and the temperature of the asphalt will be further increased through heat transfer, increasing the fluidity of the asphalt. After a long period of immersion, the interfacial destruction load of aggregate and asphalt drops to a lower level and reaches a relatively balanced state, at which point the destructive effect of the water is no longer obvious [17].

To more intuitively study the relationship between aggregate heating temperature and aggregate-asphalt interface adhesion performance, this paper performs an image analysis of the aggregate destruction interface. The grayscale treatment was carried out by imagPro-Plus software, and the peeling rate of the aggregates of the four lithic properties at different heating temperatures and immersion times was calculated using the formula (1.2), as shown in Figure 15.

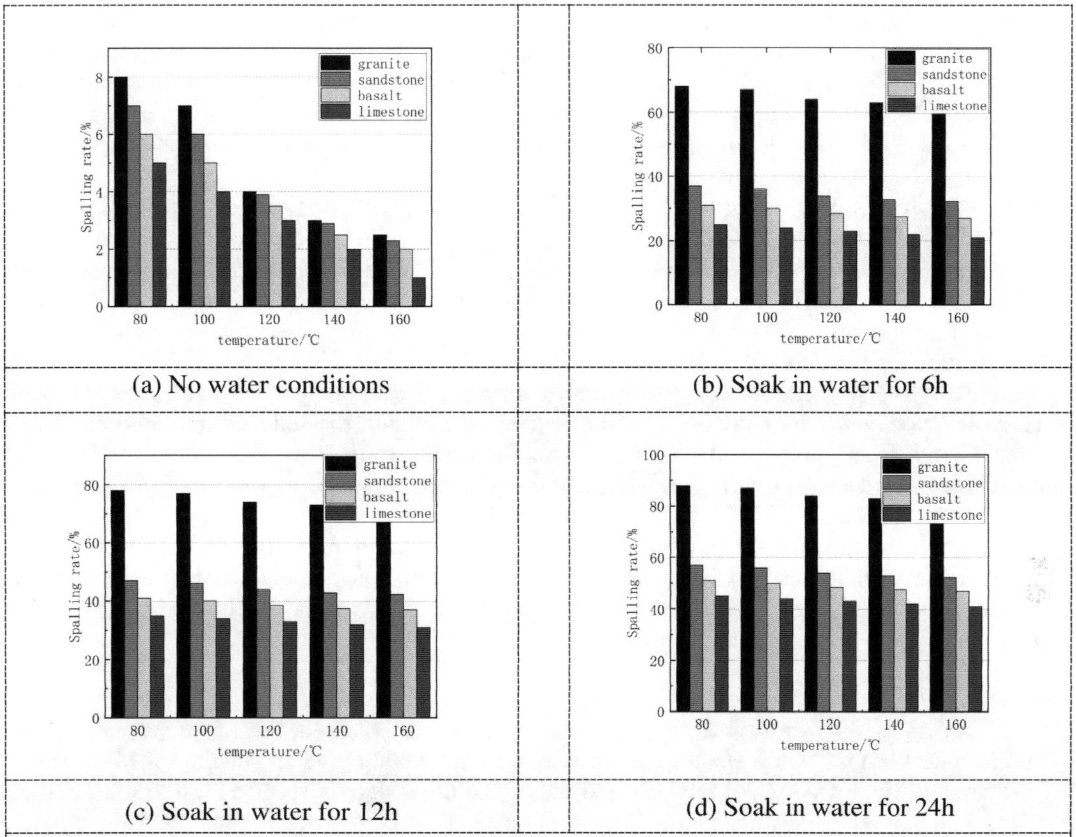

(a) No water conditions

(b) Soak in water for 6h

(c) Soak in water for 12h

(d) Soak in water for 24h

Figure 15. The interface spalling rate of each aggregate and asphalt under different soaking times and heating temperature.

It can be seen from Figure 15 that under dry conditions, the peeling rate of the aggregate and asphalt of the four lithic properties gradually decreases with the increase of the heating temperature of the aggregate, and the peeling rate is low, not more than 10%, when the heating temperature is 160 ° C, the peeling rate of the asphalt on the surface of the aggregate is only about 2%, which is mainly because the occurrence between the aggregate and the asphalt under the drying conditions is the cohesion failure, which occurs inside the asphalt, so there is less bitumen peeling on the surface of the aggregate [18]. Under immersion conditions, as the heating temperature of the aggregate increases, the peeling rate of the asphalt on the surface of the aggregate gradually decreases, mainly because with the increase of the heating temperature, the asphalt is more likely to penetrate the inside of the aggregate and is not prone to spalling. With the continuous increase of immersion time, the aggregate surface of the four lithic properties has undergone a certain degree of spalling, of which granite has the greatest degree of spalling. Under the immersion time of 24h, the spalling rate of granite under different heating temperature conditions exceeded 80%, and the aggregate and asphalt at this time had no adhesion effect, and serious adhesion damage occurred, under the action of external tension, the "sandwich" specimen was easy to break along the aggregate-asphalt interface.

4. Conclusion

The adhesion performance between asphalt and aggregate was studied by using the improved drawing test device "sandwich" specimen, and the results showed that the destructive load and sampling rate indexes could effectively evaluate the influence of water and temperature on aggregate-bitumen adhesion.

Under dry conditions, the destructive load between asphalt and aggregate gradually increases with the increase of the heating temperature of the aggregate. With the prolongation of the immersion time, the destructive load of aggregate and asphalt adhesion failure is still the largest in limestone and the lowest in granite. At the same time, as the heating temperature of the aggregate increases, the destructive load of adhesion failure between the aggregate and the asphalt gradually increases.

Under dry conditions, the peeling rate of the four lithological aggregates and asphalt gradually decreases with the increase of the heating temperature of the aggregate, and the peeling rate does not exceed 10%. When the heating temperature is 160 ° C, the peeling rate of asphalt on the surface of the aggregate is only about 2%. With the continuous increase of immersion time, the aggregate surface of the four lithic properties has undergone a certain degree of spelling, of which granite has the greatest degree of spalling. At a water immersion time of 24h, the spiraling rate of granite under different heating temperature conditions exceeded 80%.

When the immersion time is set to 6h, the interface destruction load of aggregate and asphalt gradually decreases with the increase of ambient temperature, and the reduced amplitude of the four aggregates is not much different. At the same time, the peeling rate of asphalt on the surface of each aggregate gradually decreases with the increase of ambient temperature.

Under dry conditions, the failure load and interfacial sampling rate between the four aggregates and asphalt are approximately the same. With the prolongation of the immersion time, the destructive load of aggregate and asphalt has decreased significantly, and the peeling rate of asphalt at the aggregate interface has increased significantly. When the immersion time reaches 72h, the failure load and sampling rate tend to be stable, and the aggregate and asphalt have been destroyed at this time.

References

[1] Li Xiaoyan, Bu Yin, Wang Hainian, et al. Research on Quantitative Evaluation of Morphological Characteristic of Coarse Aggregates. Journal of Building Materials, 2015, 18(03): 524-530.

[2] Wang Hainian, Hao Peiwen. Digital Description of Two-Dimensional Shape Characteristics of Coarse Aggregate. Journal of Building Materials, 2009, 12(06): 747-751.

[3] He Lu. Study on aging characteristics of SBS modified asphalt and its mixture in high temperature and rainy area. Xi'an: Chang'an University, 2019.

[4] Yu Haiyou.Study on water loss mechanism of asphalt concrete under coupling effect of water temperature and load. ChongQing: Chongqing Jiaotong University, 2014.

[5] Wang Lu. Study on phase structure and adhesion mechanism of asphalt aggregate interface. Xi'an: Chang'an University, 2014.

[6] Arabani M, Hamedi G H. Using the Surface Free Energy Method to Evaluate the Effects of Polymeric Aggregate Treatment on Moisture Damage in Hot-Mix Asphalt. Journal of Materials in Civil Engineering, 2011, 23(6): 802-811.

[7] Luo Rong, Zheng Songsong, Zhang Derun, et al. Evaluation of Adhesion Property in Asphalt-aggregate Systems Based on Surface Energy Theory. China Journal of Highway and Transport, 2017, 30(06): 209-214.

[8] Moraes R, Velasquez R, Bahia H. Using bond strength and surface energy to estimate moisture resistance of asphalt-aggregate systems. Construction & Building Materials, 2017, 130:156-170.

[9] Dou Yingying, Wei Dingbang, Li Xiaomin, et al. Adhesion Attenuation Mechanism of Asphalt-Aggregate Interface. Journal of Building Materials, 2019, 22(05): 771-779.

[10] Sun Jishu, Xu Ningqian, Li Ningli, et al. Mechanism and Application of Pitch-Based Organic Bentonite Nanocomposites. Bulletin of the Chinese Ceramic Society, 2019, 38(02): 530-536+542.

[11] Ministry of Transport of the People's Republic of China. Standard Test Methods of Bitumen and Bituminous Mixtures for Highway Engineering: JTG E20-2011. Bei Jing: China Communications Press, 2011.

[12] Ministry of Transport of the People's Republic of China. Test Methods of Aggregate for Highway Engineering: JTG E42-2005. Bei Jing: China Communications Press, 2005.

[13] Cheng Haojie, Lv Huijie, Luo Rong. Research on Geometric Characteristics of Asphalt Mixture Aggregate Based on Image Processing. Journal of Wuhan University of Technology (Transportation Science & Engineering), 2019, 43(5): 904-909.
[14] Zhang YongChang, Liu Kejia, Liang Bin. Experimental Study on Water Stability of Foamed Asphalt Mixture under Different Water Immersion Conditions. Railway Construction Technology, 2019(10): 1-4+22.
[15] Li Jiang, Li Shiqi, Yu Shengjun, et al. Optimization of Influencing Factors of Self-healing Ability of SBS Modified Asphalt Mixture. Journal of Highway and Transportation Research and Development, 2019, 36(08): 17-22.
[16] Wang Ruilin, Zhang Xiaoyan. Design Method of Warm Mix Asphalt Mixture Temperature Based on Step by Step Design. Journal of Building Materials, 2017, 20(02): 305-309.
[17] Li Bo, Chen Zhanquan, Liu Xiang, et al. Nano-adhesive Characteristics of SBS Modified Asphalt Containing Warm Mix Additive before and after Immersion in Water. Journal of Building Materials, 2018, 21(01): 60-64+84.
[18] Ji Jie, Ma Rongda, Zheng Wenhua, et al. Effect of Direct Coal Liquefaction Residue on Adhesion Characteristic between Asphalt and Aggregate. China Journal of Highway and Transport, 2018, 31(9): 27-33.

Investigation of 2D material anodes with different anions for lithium ion batteries: comparison of MoO_2, MoS_2 and $MoSe_2$

Tianrui Chen[1,*]

[1]The Experimental High School Attached to Beijing Normal University, China

*Email: chentianrui0321@163.com

Abstract. The energy storage devices used in today's society are mainly lithium batteries. At present, the anode material of commercial lithium batteries is generally graphite. Although lithium battery has superior performance compared with other energy storage methods, it still has many problems, such as poor safety, low specific capacity, and slow charging speed. In recent years, although some researchers have proposed graphene as anode material, the results show that although graphene can partly alleviate the above problems, it cannot meet the needs of industrial and domestic applications. Therefore, we investigate the properties of MoO_2, MoS_2 and $MoSe_2$ as battery anode materials. These three materials have good conductivity, and anions are the same main group. The effects of anionic electronegativity on battery performance can be compared and discussed. We obtain the binding energy, diffusion barrier, voltage, stable adsorption site and corresponding charge transfer of Li on the above three materials by first-principles calculations. We find that MoO_2, MoS_2 and $MoSe_2$ have their own advantages and disadvantages as battery anode materials. MoO_2 has the strongest binding capacity with lithium ion and the closest adsorption degree. MoS_2 is slightly inferior and $MoSe_2$ has the weakest binding capacity. This feature will lead to MoO_2 having higher specific capacity and the ability to prevent lithium dendrite growth. However, considering the lower diffusion barrier, lithium ion in $MoSe_2$ is easier to diffuse and the charging rate of the $MoSe_2$ based lithium battery would be higher, while lithium ion in MoO_2 is not easy to diffuse and the charging rate could be low. In general, the three anode can improve the battery performance in different aspects, and have a wide application prospect.

1. Introduction

With the development of technology, we have entered the information age. Electricity, as the main carrier of energy use, has become more and more important. In indoor or rail transportation, people can take electricity from the power grid through wires. However, when people go out or have no electric socket around them, they need to use a variety of batteries, including dry batteries, lead-acid batteries, liquid vanadium batteries and lithium batteries [1-2]. These batteries are widely used in mobile phones, laptops and electric vehicles. Their common feature is to convert chemical energy into electric energy [3].

In fact, the application of battery is far more than these fields. Take China as an example. According to the plan, China's carbon emissions will reach the peak in 2030, and the subsequent emissions need to be reduced year by year, which means that the utilization ratio of coal-fired power generation, that is, thermal power generation in China will gradually decline, replaced by cleaner energy power generation such as wind power generation, hydropower generation and solar power generation. Although these power generation are conducive to the sustainable development of the environment, their common problem is that they are intermittent and cannot ensure continuous and stable power supply [4-6]. Therefore, it is necessary to have a special battery to store this power: storing the power in case of excessive power generation and releasing it in case of insufficient power generation. Therefore, there can still be sufficient power supply when these power generation systems cannot work normally due to various reasons, which is enough to reflect that the battery plays an increasingly important role in life. Therefore, the research and development of battery will attract more and more attention in the future [7-8].

At present, the most widely used battery in the world is lithium battery, accounting for more than 80%. Lithium battery is usually composed of cathode, anode and electrolyte in the middle. The discharge process of the battery is that the lithium ion of the anode enters the cathode through the electrolyte, and the electrons from the external circuit to the cathode to maintain the electrical neutrality of the two electrodes. The external circuit can provide electric energy for the consumers. The charging process is opposite to the above. Driven by the external potential, lithium ions and electrons from the cathode to the anode through the internal circuit and the external circuit, respectively. The battery is reactivated through the process of charging and discharging. However, there are three main problems with the lithium batteries used today [9-11]. First, the security is poor. Taking electric vehicles as an example, if two electric vehicles are involved in collisions, a lot of heat will be generated at the damaged part. Because the electrolyte is very flammable and lithium anode has high chemical activity, the battery is easy to be ignited at high temperature and then explode, resulting in serious safety accidents. Moreover, the growth of internal lithium dendrite is also easy to lead to short circuit and ignite the electrolyte, resulting in spontaneous combustion of the battery. Second, the battery capacity is small. Note that the small capacity does not mean the small volume of the battery, but the small number of active lithium ions in the battery per unit volume or mass, so that the service time of the battery under existing given size is short. Third, the charging time is long. Take an electric vehicle as an example. If an electric vehicle needs a long-distance driving of more than 500 kilometers, it usually has one or more charges on the way. However, unlike a fuel vehicle, which can be filled with fuel in a few minutes, the charging time of an electric vehicle takes several hours to more than 10 hours, which is not conducive to long-distance driving.

To improve the performances of lithium batteries, some two-dimensional (2D) materials, such as MXene (carbides and nitrides of transition metals) [12-13] and TMDs [14-15] (transition metal dichalcogenides) are applied as the anodes. 2D materials have natural ion channels and exposed ion sites in the interlayers, which can enhance the ion diffusion rate, storage capacity and may slow down the dendrite growth. However, the effect of the anions in 2D materials on battery performances is not clear. In this paper, the possibility of MoO_2, MoS_2 and $MoSe_2$ as battery anode materials was studied. There are two reasons for choosing these three materials. Firstly, they all have good conductivity and are widely used in many fields such as optics and microelectronics [16-18]. Secondly, the anions of the three materials are the same main group, and the middle layer is Mo atom, thus the effects of anionic electronegativity on the performance of the battery can be compared and discussed by controlling a single variable.

2. Methods

We perform all the simulations based on the well-known density functional theory, as implemented in CASTEP [19] module of Materials Studio code. Exchange-correlation interactions are described by the generalized gradient approximation (GGA) within the frequently-used Perdew-Burke Ernzerhof (PBE) formulation [20]. The single electron wavefunctions are expanded to plane waves with an energy cutoff of 270 eV. In the structural optimizations, a $3 \times 3 \times 1$ k-point mesh of the Monkhorst-Pack scheme in Brillouin-zone are used with the convergence criterion of 0.05 eV/Å and 2×10^{-5} eV for ionic and electronic relaxations. To minimize the z direction periodic interaction, a >20 Å vacuum space is used. Then, the van der Waals (vdW) interactions are taken into account by the DFT-D3 method [21].

3. Results and discussions

Firstly, according to the literature, the adsorption of lithium ions by two-dimensional materials is based on van der Waals force and Coulomb force [22]. Therefore, when there are ions with high electronegativity on the surface of two-dimensional materials, it will have stronger adsorption capacity for lithium ions. In this paper, the adsorption of lithium ion by MoO_2, MoS_2 and $MoSe_2$ and their performance as anode of lithium battery are investigated. The surface atoms of these three two-dimensional materials are oxygen, sulfur and selenium. They all belong to the same main group, with similar chemical properties and certain electronegativity, so they will have a certain adsorption capacity for lithium ions. Figure 1 (a-c) are the structural diagrams of MoO_2, MoS_2 and $MoSe_2$ respectively, and the upper and lower parts are the top view and front view respectively. It can be seen that the structures of MoO_2, MoS_2 and $MoSe_2$ are very similar. They are a three-layer structure with two anions and one molybdenum ion. The two anions are on the same vertical line. From the top view, molybdenum ions are distributed in the middle of the three anions.

Figure 1 (d-f) shows the structure of the above three materials after lithium ion adsorption. It can be seen from the figure that lithium ions are not randomly distributed. The top views show the lithium ion is located at the vertex of the parallelogram (black box). The front views show the lithium ions are located on the side edges of the prism, where the parallelogram in the top view becomes a three-dimensional prism. If the lithium ion is arbitrarily placed on the two-dimensional material surface for structural optimization, the whole system will change spontaneously and the atomic position with lowest energy for the whole system will be found, so as to improve the stability of the system. Fig. 1 (g) is represented the structure of 2×2×1 supercell of Li adsorbed MoO_2. The left side is the top view and the right side is the front view. The following two new information can be seen from the figure: 1) the (x, y) coordinates of lithium ions are between the three adjacent oxygen ions, and 2) oxygen ions and lithium ions correspond one by one, that is, each oxygen ion can adsorb one lithium ion. Therefore, for single-layer MoO_2, when it is fully charged, the number of oxygen ions and lithium ions in the whole system should be equal. The lithium ion distribution of MoS_2 and $MoSe_2$ is consistent with that of MoO_2, which will not be explained separately here.

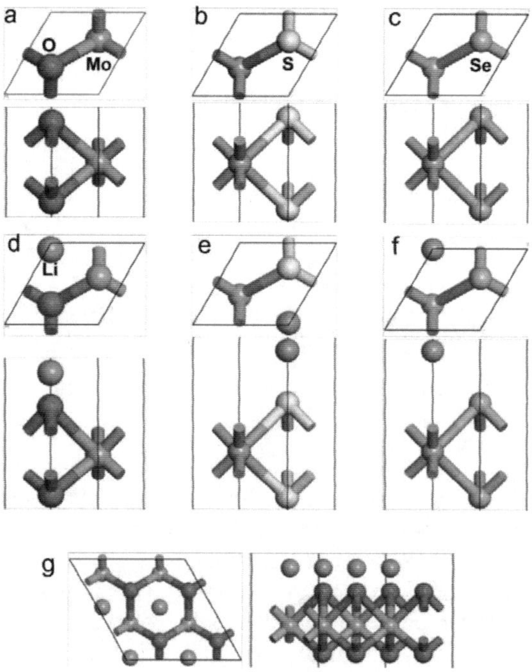

Fig. 1. Structures of (a) MoO_2, (b) MoS_2 and (c) $MoSe_2$ and the stable Li site on (d) MoO_2, (e) MoS_2 and (f) $MoSe_2$ a surfaces, and (g) is a 2×2×1 supercell of Li adsorbed MoO_2.

After understanding the most basic material structure and lithium binding sites, we further discussed the differences in the binding strength of lithium ion on the three materials. As shown below, there are two equations for calculating the average binding energy and voltage of lithium ion on electrode materials at different concentrations

$$E_B^{Li} \approx \frac{1}{x}\left[E(\text{with } xLi) - E(\text{without } xLi) - xE(Li)\right], \tag{1}$$

$$V \approx \frac{E(\text{with } x_1 Li) - E(\text{without } x_2 Li) - (x_2 - x_1)E(Li)}{(x_2 - x_1)e}, \tag{2}$$

In formula (1), E_B^{Li} refers to the average binding energy of lithium ion on other materials, E(with xLi)refers to the total energy of the material combined with lithium, E(without Li)refers to the energy of electrode materials without lithium, xE(Li) is the total energy when lithium forms the metal phase, and x is the number of lithium ions. Therefore, the average binding energy of lithium on electrode material is that the total energy of the material combined with lithium subtracts the energy of the electrode material without lithium and the total energy of lithium ions, and the final result is divided by the number of lithium ions. In equation (2), E(with x_1Li) and E(with x_2Li) is the total energy of the system before and after the increase of lithium ion concentration, $(x_2-x_1)E$ (Li) is the total energy of the metal phase composed of increased lithium, $(x_2-x_1)$is the number of increased lithium ions, and e is the charge of electrons.

According to the above formula, we calculated the changes of binding energy, charge transfer and voltage of MoO_2, MoS_2 and $MoSe_2$ with the increase of lithium ion concentration in the anode as shown in Fig. 2-4.

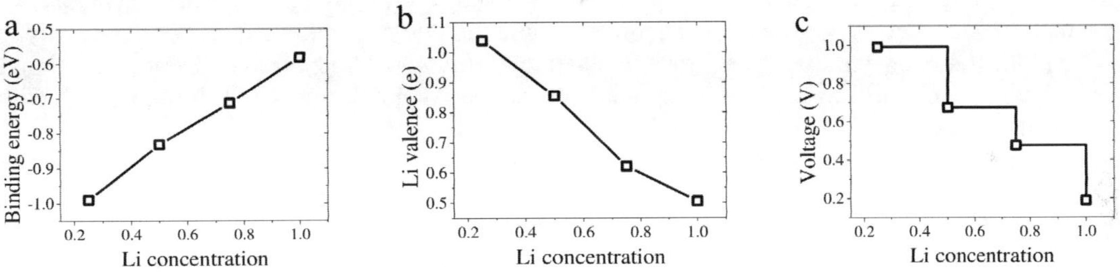

Fig. 2. (a) Binding energy of Li ion, (b) charge transfer of Li ion, and (c) voltage profile of MoO_2 anode.

Firstly, in combination with Fig. 2 (a), 3 (a) and 4 (a), it can be seen that in terms of binding energy, all of the three electrode materials increase with the lithium ion concentration. The increase of binding energy indicates that if new lithium ions enter, other ions will repel the entered lithium ions, making it more difficult for lithium ions to enter the material. The higher the lithium ion concentration, the worse the stability of lithium ions combined with anode material. Compared with the binding energy of the three materials at the same concentration, that of $MoSe_2$ is always the highest and the stability is the worst. The binding energy of MoS_2 is lower than $MoSe_2$ but higher than MoO_2. The binding energy of MoO_2 is always the lowest among the three materials at all ion concentrations, so the adsorption capacity for lithium ions is the strongest. Stronger adsorption capacity of lithium ions can facilitate a high capacity, and prevent the formation of lithium dendritic, which is conducive to the improvement of battery performance.

Secondly, in terms of electron transfer (see 2 (b), 3 (b) and 4 (b)), the average charge transfer amount from Li to MoO_2, MoS_2 and $MoSe_2$ decreased with the increase of lithium ion concentration. The transfer electron drop rate of MoO_2 is the fastest when the Li ion concentration is 0.5-0.7, and MoS_2 and $MoSe_2$ were the fastest at 0.3-0.5. Then, the charge transfer amount of the three materials at the same concentration is compared. The transfer amount of $MoSe_2$ is always the lowest, the transfer amount of MoS_2 is medium, and the MoO_2 is the highest, which can also show that the contact between lithium ion and MoO_2 is always the closest, MoS_2 is the second, and $MoSe_2$ is the least.

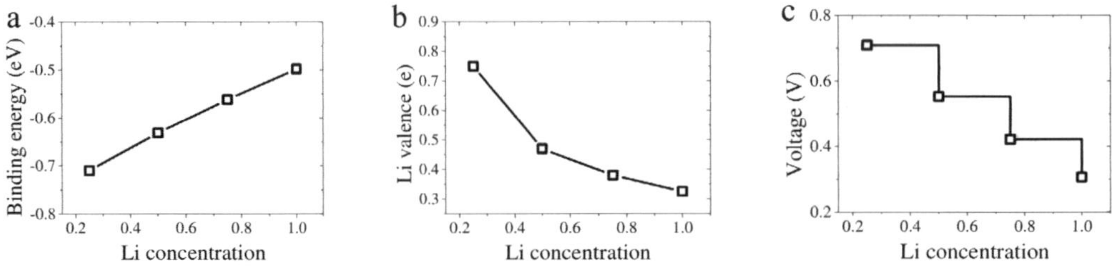

Fig. 3. (a) Binding energy of Li ion, (b) charge transfer of Li ion, and (c) voltage profile of MoS_2 anode.

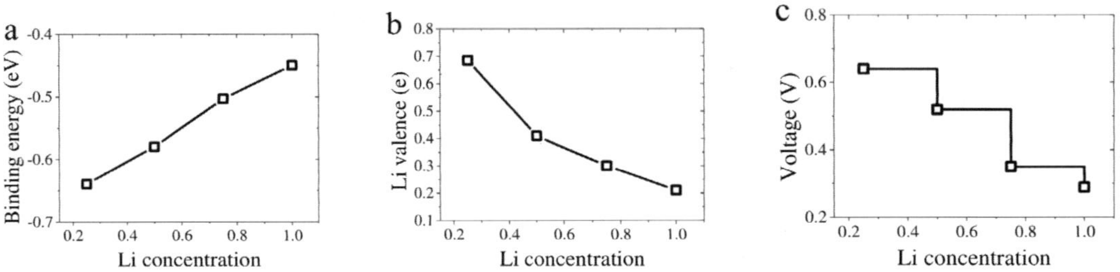

Fig. 4. (a) Binding energy of Li ion, (b) charge transfer of Li ion, and (c) voltage profile of $MoSe_2$ anode.

Finally, in terms of voltage, as shown in Fig. 2 (c), 3 (c) and 4 (c), the three materials decreased step by step with the increase of lithium ion concentration, and the rate of decline values were basically the same. The phenomenon indicates that for each part of lithium ion increased, the binding strength between the new lithium ion and the electrode was weaker than the previous lithium ion. When comparing the voltages of the three materials at the same concentration, that of $MoSe_2$ is always the lowest, the middle is MoS_2, and MoO_2 is always the highest. From these results, we can confirm that the adsorption capacity and contact tightness of lithium ions in the three materials are $MoO_2 > MoS_2 > MoSe_2$.

In the above results, the changes of binding energy, charge transfer amount and voltage of lithium ion on three materials with lithium ion concentration are discussed. Here, the charge transfer is intuitively reflected by the charge isosurface. Fig. 5 (a-c) is a top view and a front view of the electron transfer changes of MoO_2, MoS_2 and $MoSe_2$ with the increase of lithium ion concentration, respectively. Where yellow represents loss of charge and blue represents gain of charge. We can see that in these figures, lithium and molybdenum lose electrons, while oxygen, sulfur and selenium get electrons. If we compare these figures, we can also summarize the following results: 1) by longitudinal comparison of different materials, it is not difficult to see that the coverage area of yellow and blue in Fig. 5 (a) is much larger than that in Fig. 5 (b) and Fig. 5 (c), which shows the electronegativity of oxygen is much stronger than that of sulfur and selenium and has stronger electron absorption ability. Therefore, more charge can be obtained from molybdenum ions and lithium ions. The electron absorption capacity of sulfur and selenium is weak, and the charge transfer is not as much as that of oxygen. 2) Through the horizontal comparison of Fig. 5 (a), it can be seen that when the lithium ion concentration is small, the electron distribution transferred on the oxygen ion presents a hill shape and irregular shape. If the lithium ion MoO_2 is filled, that is, each oxygen ion corresponds to a lithium ion, the electron distribution transferred by the oxygen ion presents a regular sphere. If the lithium ion is not filled, the shape of the electron distribution will be biased to the side where the lithium ion exists, that is, the charge transfers around lithium ion positions. Through the above phenomena, we can clearly see the charge transfer of lithium ion to oxygen ion. 3) For the horizontal comparison of Fig. 5 (b-c), when looking at the figure carefully, one can see that with the increase of lithium ion concentration, a small blue area also appears on molybdenum ion, which shows that molybdenum ion also obtains electrons. After our analysis, the reason why molybdenum ions get electrons has little to do with anions, but with

Journal of Physics: Conference Series **2331** (2022) 012005 doi:10.1088/1742-6596/2331/1/012005

lithium. In fact, there is a displacement reaction between molybdenum and lithium. Because lithium is an alkali metal with strong metallicity and easy to lose electrons, and the metallicity of molybdenum is less than that of lithium, the tetravalent molybdenum ions will snatch the electrons of lithium and turn into lower states. This phenomenon becomes more obvious with the increase of lithium.

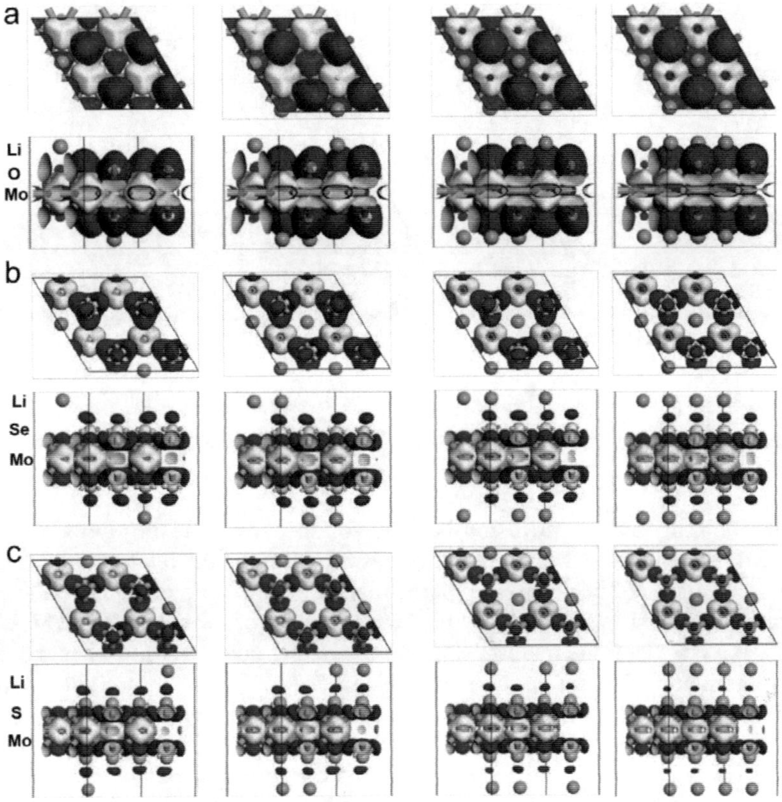

Fig. 5. Charge density of Li loaded on (a) MoO_2, (b) MoS_2 and (c) $MoSe_2$, where the blue isosurface represents charge accumulation, and the yellow isosurface represents charge consumption.

Diffusion barrier is the difference between the maximum and minimum energy of lithium ion in the whole process of lithium ion diffusion from one adsorption site to another. This diffusion process often occurs in the electrode, because lithium ions shuttle through the electrode to realize energy transfer during charge and discharge. As shown in Fig. 6 (a-c), the diffusion barriers of lithium ions on MoO_2, MoS_2 and $MoSe_2$ are shown. In the inset of the figure, we also draw the top view and front view of the diffusion path of lithium ions respectively. It is found that the total diffusion barrier of lithium ion on MoO_2 in Fig. 6 (a) is about 0.3ev. With the change of lithium ion position, its energy get the highest values at points 2 and 4. Different from Fig. 6 (b) and (c), the energies of lithium ion on MoS_2 and $MoSe_2$ present the maximums at 3 points. If we compare the diffusion barriers of lithium ions on the three materials, we can find that MoO_2 and MoS_2 are about 0.3ev, and $MoSe_2$ was significantly lower than them, only 0.17 eV. According to these data, we conclude that the diffusion barrier of lithium ion on $MoSe_2$ is the lowest and easier to diffuse, so the charge/discharge rate will increase significantly. The diffusion barriers of MoO_2 and MoS_2 are high, so the improvement of the charging rate will not be as obvious as $MoSe_2$.

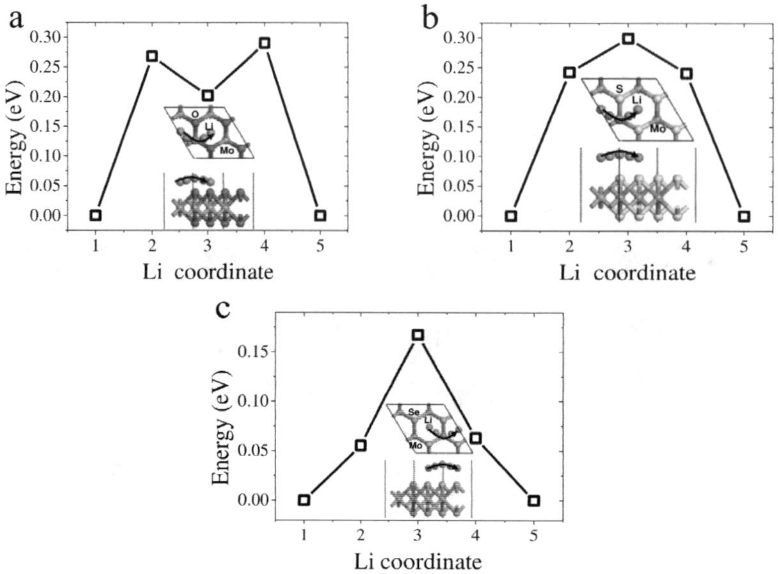

Fig. 6. Energy profiles of Li diffusion on (a) MoO_2, (b) MoS_2 and (c) $MoSe_2$ surface.

4. Conclusions

In this study, the binding energy, diffusion barrier, voltage, stable adsorption site and corresponding charge transfer of Li in MoO_2, MoS_2 and $MoSe_2$ were calculated by first principles method. The effects of anion electronegativity on the performance of two-dimensional material battery were discussed. According to the results, MoO_2, MoS_2 and $MoSe_2$ have their own advantages and disadvantages as battery electrode materials. Firstly, we conclude that MoO_2 has the strongest binding capacity with lithium ion, the closest adsorption degree, MoS_2 is in the middle, and $MoSe_2$ has the weakest binding capacity. Therefore, the actual capacity of MoO_2 in the battery composed of equal molar amount of MoO_2, MoS_2 and $MoSe_2$ will be the largest, and MoO_2 cell is also the least likely to form dendrites. However, results show that $MoSe_2$ has the lowest diffusion barrier, i.e. lithium ion is easier to diffuse and the charging rate is higher, while lithium ion in MoO_2 is not. In conclusion, the above three materials have their own advantages, which can further solve the problems of battery safety, capacity and charging rate, and are expected to become promise anode materials of the next generation of batteries. Therefore, when using these three materials, they can be selected according to the application requirements and their characteristics. In this study, this material only stays in the theoretical stage. If we want to make the battery composed of the above three materials really enter the market, we also need to carry out physical safety certification, find their simple generation conditions, and have the ability of mass production.

References

[1] Ki Jae Kim, Min-Sik Park, Young-Jun Kim, Jung Ho Kim, Shi Xue Doub, M. Skyllas-Kazacos (2015). A technology review of electrodes and reaction mechanisms in vanadium redox flow batteries. Journal of Materials Chemistry, 3(33), 16913-16933.

[2] Nitta, N. , Wu, F. , Lee, J. T. , & Yushin, G. . (2015). Li-ion battery materials: present and future. Materials Today, 18(5) 252-264.

[3] Kurzweil, P. . (2015). Lithium battery energy storage. Electrochemical Energy Storage for Renewable Sources and Grid Balancing, 269-307.

[4] Haines, A. , Smith, K. R. , D Anderson, Epstein, P. R. , Mcmichael, A. J. , & Roberts, I. , et al. (2007). Policies for accelerating access to clean energy, improving health, advancing development, and mitigating climate change. Lancet, 370(9594), 1264-1281.

[5] Kamat, P. V. . (2007). Meeting the clean energy demand: nanostructure architectures for solar energy conversion. Journal of Physical Chemistry C, 111(7), 2834-2860.
[6] Su, P. , Liu, Y. , Zhang, J. , Chen, C. , & Zhao, X. . (2020). Pb-based perovskite solar cells and the underlying pollution behind clean energy: dynamic leaching of toxic substances from discarded perovskite solar cells. Journal of Physical Chemistry Letters, 11(8), 2812-2817.
[7] Qiao, Y. , Yang, H. , Chang, Z. , Deng, H. , & Zhou, H. . (2021). A high-energy-density and long-life initial-anode-free lithium battery enabled by a li2o sacrificial agent. Nature Energy, 6(6), 1-10.
[8] Zhao, E. , Zhang, Z. G. , Li, X. , He, L. , Yu, X. , & Li, H. , et al. (2020). Neutron-based characterization techniques for lithium-ion battery research. Chinese Physics B, 29(1), 018201 (22pp).
[9] Zeng, X. , Li, J. , & Liu, L. . (2015). Solving spent lithium-ion battery problems in china: opportunities and challenges. Renewable & Sustainable Energy Reviews, 52, 1759-1767.
[10] Huang, Y. , Zhang, Z. , Gao, H. , Huang, J. , & Li, C. . (2020). Li1.5al0.5ti1.5(po4)3 enhanced polyethylene oxide polymer electrolyte for all-solid-state lithium batteries. Solid State Ionics, 356, 115437.
[11] Li, L. , Wang, C. , Yan, S. , & Zhao, W. . (2020). A combination state of charge estimation method for ternary polymer lithium battery considering temperature influence. Journal of Power Sources, 484, 229204.
[12] Xiao Tang,Xin Guo,Wenjian Wu,Guoxiu Wang, 2D Metal Carbides and Nitrides (MXenes) as High-Performance Electrode Materials for Lithium-Based Batteries, Advanced Energy Materials, 26, 2018, 1810897.
[13] Qiwei Pan, Yongwei Zheng, Sankalp Kota, Weichun Huang, Shijun Wang, Hao Qi, Seyong Kim, Yingfeng Tu, Michel W. Barsoum, Christopher Y. Li, 2D MXene-containing polymer electrolytes for all-solid-state lithium metal batteries, 2019,1, 395-402
[14] Tyler Stephenson, Zhi Li, Brian Olsen, David Mitlin, Lithium ion battery applications of molybdenum disulfide (MoS2) nanocomposites, 2014,7, 209-231
[15] Liang Zhang, Dan Sun, Qiulong Wei, Huanxin Ju, Jun Feng, Junfa Zhu, Liqiang Mai, Elton J Cairns, Jinghua Guo, Understanding the electrochemical reaction mechanism of VS2 nanosheets in lithium-ion cells by multiple in situ and ex situ x-ray spectroscopy, 2018 J. Phys. D: Appl. Phys. 51 494001.
[16] Lee, H. S. , Min, S. W. , Chang, Y. G. , Park, M. K. , Nam, T. , & Kim, H. , et al. (2012). Mos2 nanosheet phototransistors with thickness-modulated optical energy gap. Nano Letters, 12(7), 3695-3700.
[17] A Molina-Sánchez, Sangalli, D. , Hummer, K. , Marini, A. , & Wirtz, L. . (2013). Effect of spin-orbit interaction on the optical spectra of single-layer, double-layer, and bulk mos2. Physical review. B, Condensed matter, 88(4), 2358-2367.
[18] Liao, W. , Wei, W. , Tong, Y. , Chim, W. , & Zhu, C. . (2017). Electrical performance and low frequency noise in hexagonal boron nitride encapsulated mose2 dual-gated field effect transistors. Applied Physics Letters, 111(8), 082105.
[19] Clark, Stewart J., Segall, Matthew D., Pickard, Chris J., Hasnip, Phil J., Probert, Matt I. J., Refson, Keith and Payne, Mike C.. "First principles methods using CASTEP" Zeitschrift für Kristallographie - Crystalline Materials, vol. 220, no. 5-6, 2005, pp. 567-570. https://doi.org/10.1524/zkri.220.5.567.65075
[20] J. P. Perdew, K. Burke, and M. Ernzerhof, Generalized gradient approximation made simple, Phys. Rev. Lett. 77, 3865 (1996).
[21] G. Stefan, E. Stephan, and G. Lars, Effect of the damping function in dispersion corrected density functional theory, J. Comput. Chem. 32, 1456 (2011).
[22] Aierken, Yierpan, Sevik, Cem, Gulseren, & Oguz, et al. (2018). Mxenes/graphene heterostructures for li battery applications: a first principles study. Journal of Materials Chemistry, A. Materials for energy and sustainability, 6(5), 2337-2345.

Investigation and Improvement of Switching Characteristics of SiC Optically Controlled Transistor

Yingying Zhang[1,2]

1. Department of Electrical Engineering, Sanmenxia Polytechnic, Sanmenxia 472000, China
2. College of Applied Engineering, Henan University of Science and Technology, Luoyang 471023, China

yht197979@163.com

Abstract. Aiming to the switching characteristics of SiC optically controlled transistor, SiC NPN optically controlled transistor was investigated through Silvaco TCAD. The results show that under 4500V bias voltage, the turn-on and turn-off dV/dt of the SiC transistor are 428.5V/ns and 23.9V/ns, respectively. And the tailing problem in the turn-off process is obvious. In order to improve the switching characteristics of SiC optically controlled transistor, the minority carrier lifetime in base layer is regional controlled. The simulation results indicate that, by using minority carrier lifetime control technology, the turn-off time and turn-off dV/dt are improved by about 28.2% and 39.3%, respectively. Meanwhile, turn-on time and turn-on dV/dt are only degenerated by about 3.6% and 3.4%, respectively. The overall level of switching characteristics of SiC optically controlled transistor are improved.

1. Introduction

Because of the high peak power output, pulsed power technology has gained important applications in industrial fields[1,2]. As the core component of pulse power system, the characteristics of pulse power switch affect the performance level of pulse power system directly. In recent years, with the development of SiC power semiconductor device technology, SiC pulse power switching devices with higher voltage resistance, faster switching speed, stronger temperature resistance and smaller volume have become the focus of research in the field of pulse power switching technology[3,4]. Compared with electrically controlled power devices, SiC light triggered thyristor (LTT), SiC photo-conductive switch (PCSS), SiC optically controlled transistor (OCT) and other optically controlled pulse power switching devices have unique advantages in simplified driving circuit and improved anti-electromagnetic interference ability[5-7]. Which further improve the application potential of optically controlled power devices in national defense and industrial fields[5,7]. Compared with SiC PCSS, SiC LTT and SiC OCT have the advantages of low leakage and high gain, and can achieve lower power loss as a pulse power switch[6-8]. Compared with SiC LTT, SiC OCT has controllable turn-off characteristics and can be used as a pulse power switch to achieve a high repetition rate pulse switch[6,8].

In this paper, the switching characteristics of SiC NPN optically controlled transistor are investigated. In order to improve the switching characteristics of SiC optically controlled transistor, optimization design of minority carrier lifetime in base region is carried out and simulated.

Journal of Physics: Conference Series **2331** (2022) 012006 doi:10.1088/1742-6596/2331/1/012006

2. Device structure and static characteristics simulation

Figure 1 shows the half-cell profile structure of SiC NPN optically controlled transistor. As shown in the figure, the width of the half-cell is 10μm and a width of emitter layer is 4.0μm. In SiC NPN optically controlled transistor, from top to bottom are n-type emitter region, p-type base region, n-type collector region, n-type buffer layer and n-type substrate. The thickness of each layer in SiC NPN optically controlled transistor are 1.0μm, 2.5μm, 60μm, 2.0μm, and 100μm, respectively. The doping concentration of emitter region, base region and collector region are $2.0\times10^{19}cm^{-3}$, $5.0\times10^{17}cm^{-3}$, and $2.0\times10^{14}cm^{-3}$, respectively.

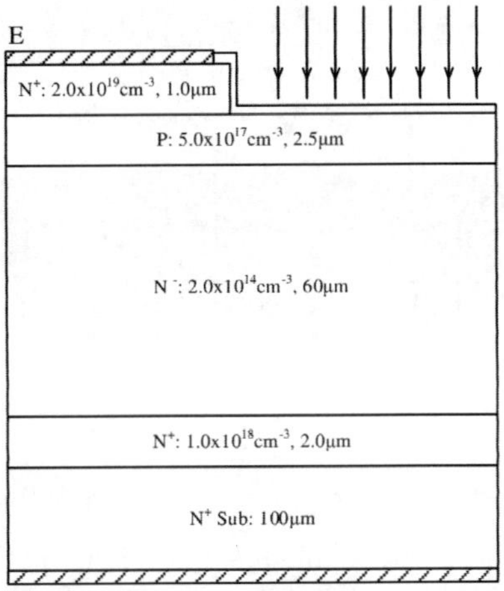

Figure 1. Half-cell profile structure of SiC NPN optically controlled transistor.

The static characteristics of SiC NPN optically controlled transistor are simulated by Silvaco TCAD. The simulated curves of output characteristic are shown in Figure 2(a). The corresponding minority carrier lifetime in the base region is 50ns. As can be seen from the figure, when the voltage between collector and emitter (V_{CE}) is 5V and ultraviolet (UV) light intensity is increased from $1W/cm^2$ to $1000W/cm^2$, collector current density (J_C) is increased from $1.1A/cm^2$ to $205A/cm^2$. The Early effect becomes more obvious as the intensity of ultraviolet light increases. It is caused by the photo-conductance of ultraviolet light to the base region of SiC optically controlled transistor. Figure 2(b) shows the breakdown characteristic curve between collector and emitter of SiC optically controlled transistor. As shown in the figure, the breakdown voltage of the transistor is 9465V. The corresponding leakage current density is $10\mu A/cm^2$.

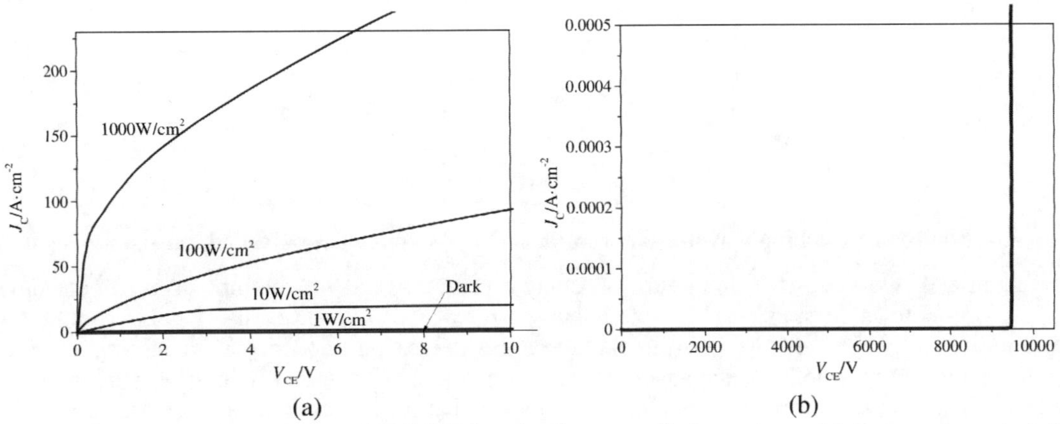

Figure 2. Static characteristics of SiC NPN optically controlled transistor. (a) Output characteristic curves. (b) Breakdown characteristic curves.

3. Switching characteristics investigation and improve

During switching process, the current change rate (di/dt) of the device is affected by the inductance of the external circuit. In order to investigate the switching characteristics of SiC optically controlled transistor, the voltage change rate (dV/dt) is used to discuss the switching characteristics of SiC optically controlled transistor. The switching characteristics of SiC optically controlled transistor are simulated in a pure resistive load circuit through Silvaco TCAD. During the simulation, 365nm UV light with 50ns pulse width and 1kW/cm^2 light intensity is used as base control signal. Schematic diagram of the circuit is shown in Figure 3. In the circuit, the load resistance R_L is 10Ω, the power supply U_S is 4500V, and the capacitor C_S is 200μF.

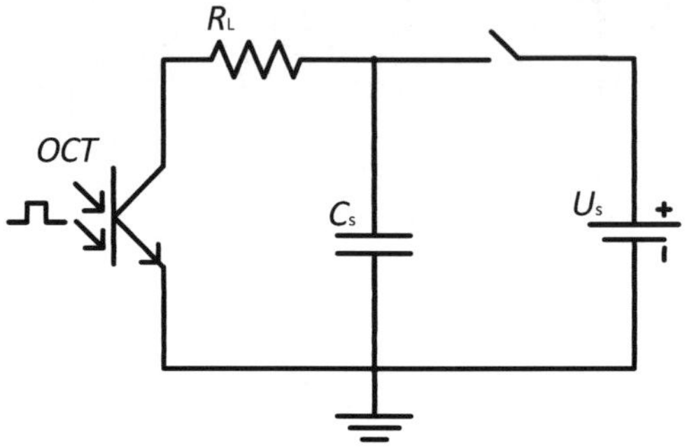

Figure 3. Schematic diagram of the pure resistive load circuit.

Switching characteristics of SiC NPN optically controlled transistor is shown in figure 4. The corresponding minority carrier lifetime in the base region is 50ns. As shown in the figure, the peak current of the transistor is 445.6A/cm^2. Turn-on time (t_{on}) and turn-off time (t_{off}) of the transistor are 8.4ns and 150.7ns, respectively. Turn-on delay time ($t_{d,on}$) and turn-off delay time ($t_{d,off}$) of the transistor are 0.5ns and 118ns, respectively. Under control of 1kW/cm^2 UV light, the turn-on dV/dt and turn-off dV/dt are 428.5V/ns and 23.9V/ns, respectively. The small turn-off dV/dt of the transistor is mainly caused by the long turn-off delay. The SiC NPN optically controlled transistor shows obvious trailing problem, which limits the switching frequency of the SiC optically controlled transistor.

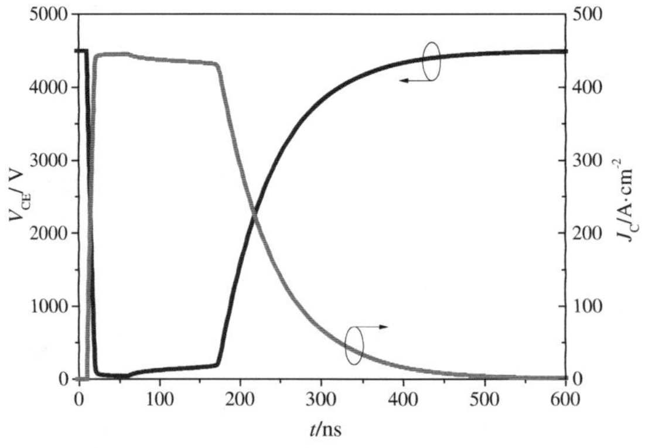

Figure 4. Switching waveform curves of SiC NPN optically controlled transistor.

Effect of UV light intensity (I_{uv}) on turn-off delay time, turn-on dV/dt and turn-off dV/dt are shown in Figure 5. With increasing of UV light intensity, turn-on dV/dt and turn-off delay time exhibit increasing tendency. Meanwhile, turn-off dV/dt exhibit decreasing tendency with increasing of UV light intensity. On the whole, switching-on performance is improved and switching-off performance is degenerated with UV light intensity increasing. It means that the switching characteristic of SiC NPN optically controlled transistor can not be optimized through increasing UV light intensity.

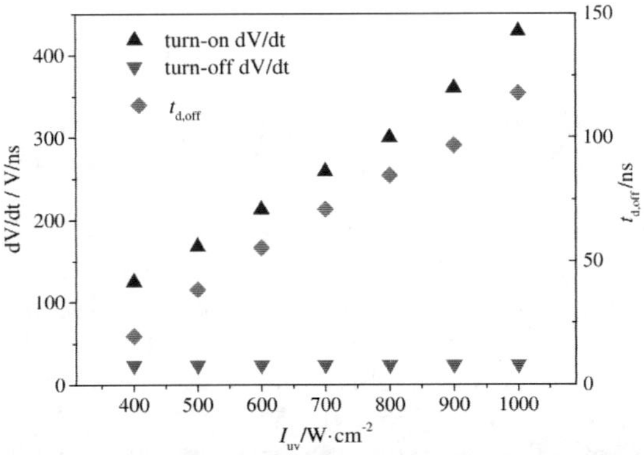

Figure 5. Effect of UV light intensity on switching characteristic of SiC transistor.

In order to improve the switching characteristics of SiC NPN optically controlled transistor, keeping balance between turn-on and turn-off processes is important. As shown in figure 6(a), minority carrier lifetime in base region is optimal designed. The minority carrier lifetime in 0.5μm thick base layer near collector junction is shorten to be 5.0ns, while the minority carrier lifetime in upper region of base layer is still 50ns. Figure 6(b) shows the simulated switching waveforms of the optimal designed transistor, in which the dotted line is the simulated switching waveforms of the conventional transistor from figure 4.

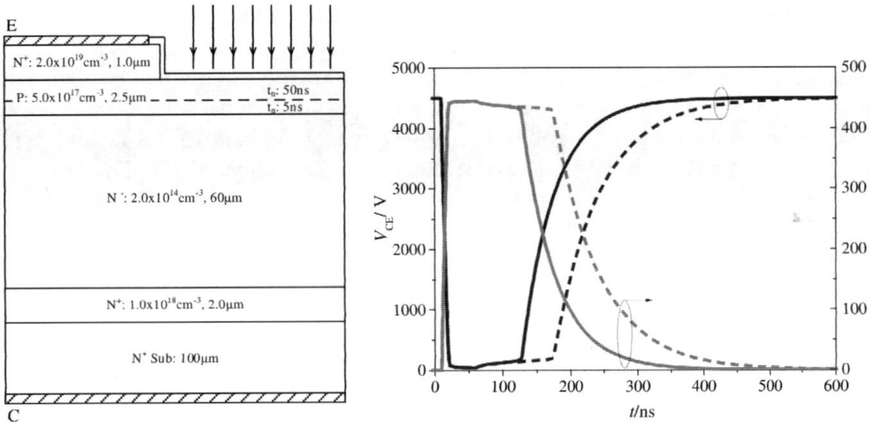

(a) (b)

Figure 6. Optimal designed SiC NPN optically controlled transistor. (a) Structure diagram. (b) Switching waveforms.

As shown in the figure 6(b), the peak current of the transistor is same. While the turn-off process of the optimal designed transistor is obviously improved compared to conventional one. By optimal design of the minority carrier lifetime, the turn-on time and turn-off time of the transistor are changed to be 8.7ns and 108.2ns, respectively. Turn-on delay time and turn-off delay time are changed to be 0.5ns and 70.3ns, respectively. Turn-on dV/dt and turn-off dV/dt are changed to be 413.8V/ns and 33.3V/ns, respectively. Compared with conventional transistor, the turn-off time, turn-off delay time and turn-off dV/dt are improved by about 28.2%, 40.4% and 39.3%, respectively. While the turn-on time and turn-on dV/dt of optimal designed transistor are only degenerated by about 3.6% and 3.4%, respectively. It is because that during the initial stage of switching-on process, the lower region of the base layer is exhausted, the minority carriers is transported by drift mechanism. Shortening of the minority carrier lifetime in the lower region of the base layer has little effect on the turn-on characteristic of the transistor. While shortening of the minority carrier lifetime in the lower region of the base layer can also accelerate the recombination of the minority carriers in the base region, when during turn-off process. As a result, shortening of the minority carrier lifetime in the lower region of the base layer is effective in improving the switching characteristic of the SiC NPN optically controlled transistor.

4. Conclusion

Switching characteristics of SiC NPN optically controlled transistor are investigated in this paper. The simulation results indicate that the SiC NPN optically controlled transistor has obvious trailing problem. And the switching characteristics of the transistor can not be improved by increasing UV light intensity. In order to improve the switching characteristics, the SiC transistor is optimal designed through regional control of minority carrier lifetime in base layer. By using minority carrier lifetime control technology, the turn-off time, turn-off delay time and turn-off dV/dt are improved by about 28.2%, 40.4% and 39.3%, respectively. While the turn-on time and turn-on dV/dt of optimal designed transistor are only degenerated by about 3.6% and 3.4%, respectively. The overall level of switching characteristics of SiC NPN optically controlled transistor are improved. The research results provide an important technical reference for the application of SiC optically controlled transistor in pulse power switching technology.

5. References

[1] Wei J, Li Z, Li B. Investigation of Reverse Recovery Current of High-Power Thyristor in Pulsed Power Supply. Electronics. 2020, 9(8):1292.

[2] C Zeng, L F Deng, Z J Li, J Wang. Experimental comparison of SiC GTO and ETO for pulse power applications. J. Semicond., 2018, 39(12): 124017.

[3] M. Agamy, F. Tao and A. Elasser. High-Speed Medium-Voltage SiC Thyristors for Pulsed Power Applications. IEEE Transactions on Industry Applications, 57(4):3812.

[4] M. Azizi, J. J. van Oorschot and T. Huiskamp. Ultrafast Switching of SiC MOSFETs for High-Voltage Pulsed-Power Circuits. IEEE Transactions on Plasma Science, 48(12):4262.

[5] T. G. Subhash Joshi and V. John. Performance Comparison of ETT- and LTT-Based Pulse Power Crowbar Switch. IEEE Transactions on Plasma Science, 45(11):2994.

[6] F. Zhao and M. M. Islam. Optically Activated SiC Power Transistors for Pulsed-Power Application. IEEE Electron Device Letters, 31(10):1146.

[7] X. Chu et al.. MHz Repetition Frequency, Hundreds Kilowatt, and Sub-Nanosecond Agile Pulse Generation Based on Linear 4H-SiC Photoconductive Semiconductor. IEEE Transactions on Electron Devices, 69(2):597.

[8] X. Wang, H. Pu, Q. Liu, L. An, X. Tang and Z. Chen. Demonstration of 4H-SiC Thyristor Triggered by 100-mW/cm^2 UV Light. IEEE Electron Device Letters, 41(6):824.

Preparation and investigation indium tin oxide for transparentelectrode of MS Schottky photodiode

N Sangwaranatee[1] and I Srithanachai[2]

[1]Applied Physics Research Group, Faculty of Science and Technology, Suan Sunandha Rajabhat University, Bangkok 10300, Thailand
[2]Department of Electronics, Faculty of Engineering, King Mongkut's Institute of Technology Ladkrabang, Bangkok 10520, Thailand

Corresponding author: narong.sa@ssru.ac.th

Abstract. In this paper investigation photocurrent of MS Schottky photodiode by using transparent electrode. The transparent material use in this investigation is indium tin oxide (ITO). Transparent electrode can help to get more photocurrent under contact due to light can penetrate though into depletion region then generate photocurrent. ITO will prepare by RF sputtering technique and investigate material characteristics such as electrical and optical properties. Optimize ITO properties after various thickness by growth sputter time show high transparent around 91.2% with low resistance $3.4x10^{-3}$ ohm-cm. Build in voltage of device show 0.3V by same with metal contact and current under light exposure show around 1.5mA at 25,000 lux with bias -5V.

1. Introduction

Photodiode is one of device use for many facility. Photodiode use for light measurement in science and industry work by detector fabrication from semiconductor material such as silicon [1], InGaAs [2] and GaN [3]. Currently, photodiode can separate for many structure that we use in our application such as photoconductor [4], P-N photodiode [5], P-i-N photodiode [6], avalanche photodiode [7] and Schottky photodiode [8]. In this paper will investigation ITO properties and apply for transparent electrode of Schottky photodiode. Schottky photodiode in one type of semiconductor diode that can use for switching and photodetector [9]. Schottky diode use low energy with high switching speeds. Structure of Schottky diode consist of metal and semiconductor. Thin metal growth in active area and light can penetrate though thin metal for generate electron-hold pare in depletion region when reverse bias. Metal contact has limit in light absorption range due to light cannot penetrate metal if thick, bad uniformity and high reflection. Moreover, thin metal is hard to control in fabrication process. One of material can use in active area of MS Schottky is transparent material [10]. Figure 1 show Schottky photodiode with metal contact by the metal contact prepare with thin thickness for light penetrate. Although, we use metal for long time but still has some limit in process to control thickness and may reduce device performance if thickness and uniformity error. [11]

Indium tin oxide (ITO) can use for photodiode contact because high transparent and low resistance. ITO is world wide material use in many application such as solar cell, defrosting window and flat panel display. ITO has cubic bixbyite structure by doped Sn into indium oxide (InO) structure. [12] Goal of this paper will investigate ITO thin film that optimize for photodetector transparent electrode and fabricate ITO to MS Schottky photodiode.

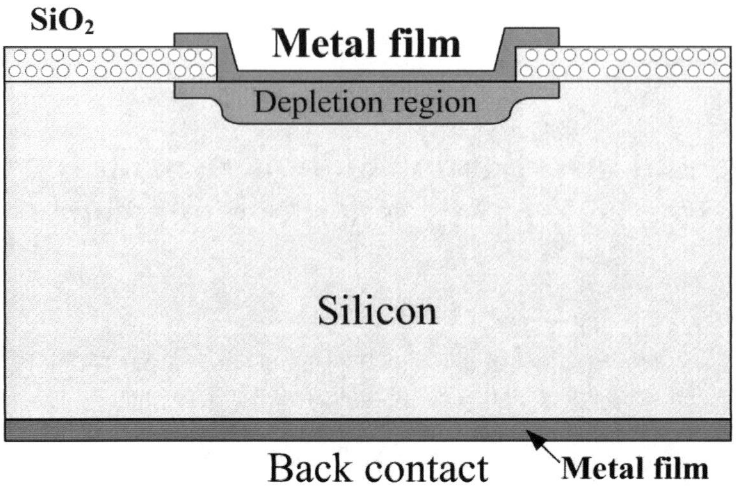

Figure 1. MS Schottky photodiode with metal in active area.

2. Experiment procedure

Indium tin oxide (ITO) preparation process by RF sputter with 2 different thickness 580 and 1,180 nm. ITO this film will measure transparent (%T) and resistivity (resistivity).

Device fabrication under CMOS technology process by follow below: [13]

Silicon substate cleaning process
Thermal oxidation 1
Doping process (n^+) backside of silicon substrate
Oxidation 2
Photolithography 1
Metal coating backside by Vacuum evaporation
Photolithography 2
RF sputtering ITO
Photolithography 3
Sintering process under 300 and 480 °C in Nitrogen environment

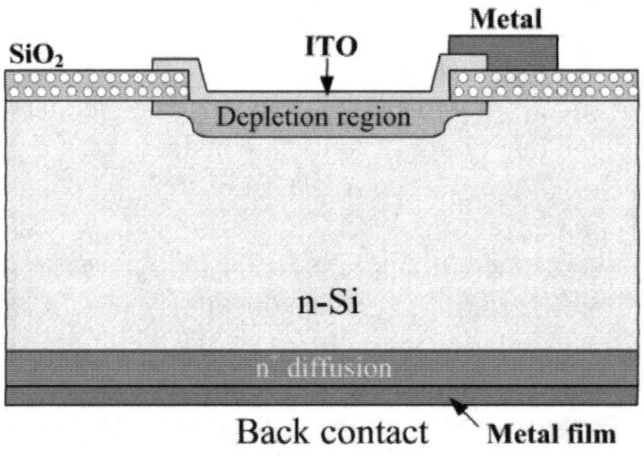

Figure 2. MS Schottky photodiode with ITO thin film in active area.

Journal of Physics: Conference Series **2331** (2022) 012007 doi:10.1088/1742-6596/2331/1/012007

Figure 2. show device after fabrication by using ITO in active area and growth metal for wire contact. Electrical properties will measure under dark environment and exposure light 5,000-25,000 lux with reverse sweep 0 to -5V.

3. Results and Discussions

ITO this film after prepared by RF sputtering show in figure 3. Growth rate of RF sputter that we use prepare ITO thin film around 20nm/min and various thickness 580 and 1,180 nm.

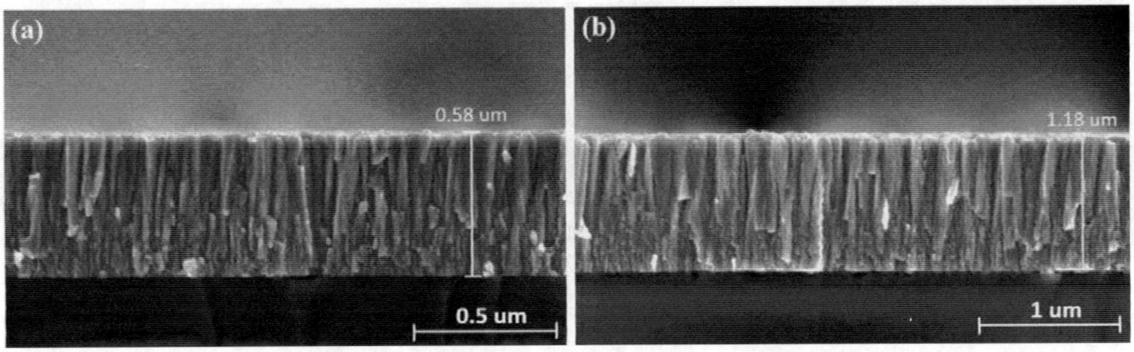

Figure 3. ITO thin film thickness (a) 580 nm and (b) 1,180 nm.

Transparent of thin film show in figure 4, results show transparent in wave range 300-1000 nm by highest transparent (%T) of this film around 97.3% with 1180 nm. Thin film thickness 1,180 show high transmission due to film show good uniformity and low reflection.

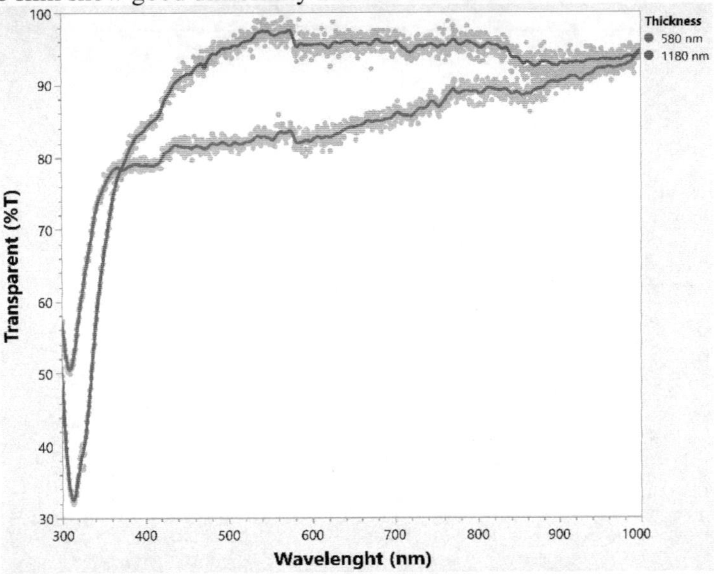

Figure 4. Transparent of ITO thin film.

Transparent and resistivity of ITO thin film summary show in table 1. From the table show film thickness 1,180 nm has low resistivity and high transparent properties. The resistivity relates with film uniformity and crystallization. Light will scatter and reflect if thin film has low uniformity [14]. However, thickness to high will impact for transmission properties because light may scatter from grain of thin film. [15]

Table 1. Summary of transparent and resistivity of ITO thin film.

Thickness (nm)	580	1,180
Transparent (%) @700nm	82.4	91.2
Resistivity (10^{-3} ohm-cm)	5.6	3.4

Figure 5 show the current-voltage characteristics of ITO/n-Si Schottky diode shows build in voltage (V_{bi}) around 0.3V. Dark current of device show around $2x10^{-7}$ Ω-cm.

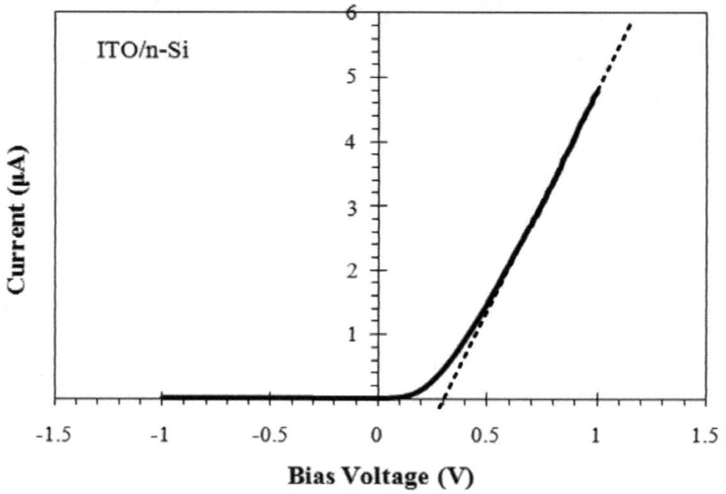

Figure 5. Current-voltage characteristics of MS Schottky photodiode with ITO thin film.

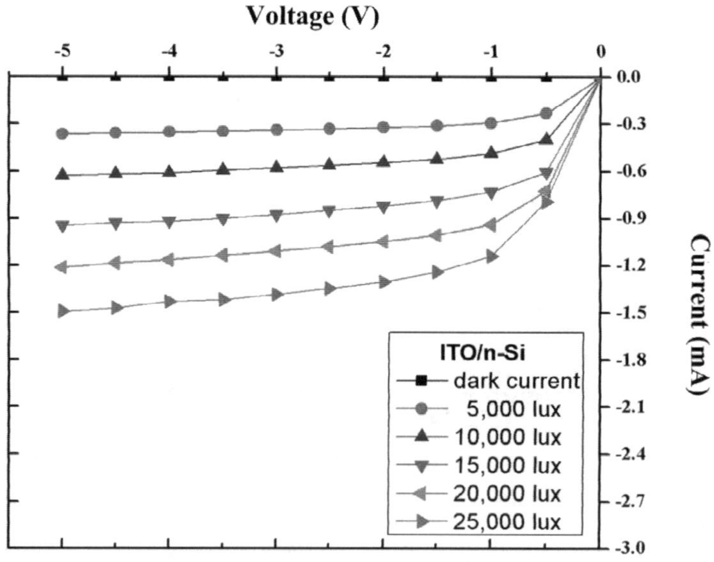

Figure 6. Photocurrent of device while exposure various light 5,000-25,000 lux.

Figure 6 shows current of photodetector by various light density 5,000-25,000 lux with voltage sweep 0 to -5V. The current show highest current around 1.5 mA at 25,000 lux and -5V.

4. Conclusion

The goal of this paper will investigate ITO thin film for transparent electrode of MS Schottky photodiode. From the results show thin film transparent around 91.2% at light wavelength 700 nm by thin film thick 1180 nm. Thin film show good uniformity with thickness 1,180 nm that will reduce light scatter and reflection. Resistivity of 1,180 nm thick film not high and good for transparent electrode. MS Schottky after fabrication process by using ITO thin film for electrode show build in voltage same with metal electrode and current after exposure various light intensity also show good. From this investigation show that ITO thin oxide can use for MS Schottky electrode although current while exposure light not significant high that may from active area quite small. In next investigation, we will investigate photocurrent by various active area size.

References

[1] Sebastian L, Ray T C and Chennupati J 2018 *Semicond. Semimet.* **99** 2
[2] Ajayan J and Nirmal D 2015 *Superlattices Microstruct.* **86** 1
[3] Pearton S J, Abernathy C R, Overberg M E, Thaler G T, Onstine A H and Gila B P 2002
[4] Hitohiro M and Yuji M 1995 *J Electrostat* **35** 67
[5] Wagah F M, Mudhar A H and Munther N A T 2013 *Renew. Sust. Energ. Rev.* **26** 408
[6] Charles N P O, Helen J K and Edval J P S 2016 *Phys Med.* **32** 1495
[7] Pansart J P 1997 *Nucl. Instrum. Methods Phys Res.A: Accel. Spectrom. Detect. Assoc. Equip.* **387** 186
[8] Jiyoul L, Hwang D K, Park C H, Kim S S and Seongil I 2004 *Thin Solid Films* **451-452** 12
[9] Om P S, Alka S, Gour K S, Sudhir H and Singh V N 2016 *Sol Energy* **157** 28
[10] Hyunki K, Seung H H, Yun C P, Jaekuen L, Chil H J and Joondong K 2014 *Mater. Lett.* **115** 45
[11] Deqiang L, Zhe H, Jing Z, Qifei J and Yangyang C 2021 *Appl. Therm. Eng.* **193** 117001
[12] Micali M, Cosentino S and Terrasi A 2021 *Sol. Energy Mater Sol. Cells.* **221** 110904
[13] Narong S, Itsara S and Preecha Y 2016 *Far East J. Electron. Commun.* **16** 847
[14] Jui Y W, Chien S H, Sin L Q, Yun S C and Jung J H 2022 *Sol. Energy Mater Sol. Cells.* **234** 111433
[15] Sylvie P 2003 *European Structural Integrity Society* **31** 321

Acknowledgments

The authors would like to thanks Suan Sunandha Rajabhat University (SSRU) and King Mongkut's Institute of Technology Ladkrabang (KMITL).

Effect of Irradiation (SRFE) on MSM Photodetectors Device

Itsara Srithanachai[1] and Narong Sangwaranatee[2]

[1]Department of Electronics, Faculty of Engineering, King Mongkut's Institute of Technology Ladkrabang, Bangkok 10520, Thailand
[2]Applied Physics Research Group, Faculty of Science and Technology, Suan Sunandha Rajabhat University, Bangkok 10300, Thailand

Corresponding author: narong.sa@ssru.ac.th

Abstract. This paper presents the effect of high energy irradiation expose on metal-semiconductor-metal (MSM) photodetectors device. High energy radiation may impact to lattice structure of semiconductor device. The most impact of radiation on MSM device had been studied base on polycrystalline silicon. Based on our research in SRFE process exposed on P-N power device, radiation will interact with device mechanism and structure that change electrical properties of device. However, MSM device show electrical properties change after irradiation exposure process by photocurrent increased around 4 times when compare with non-irradiated.

1. Introduction

Semiconductor are material with conductivity control in electronics device. Semiconductor material are important for electronics and electrical area due to we use semiconductor device in many applications for our life [1]. However, semiconductor material has limit for increase the performance by many techniques use for improve device performance such as new process [2], atom doping [3] and new device structure [4]. Although, we can increase device performance but still has some challenge to reduce size of device with high performance and low require energy. [5]

In present, semiconductor device used in our life by different structure such as high way light [6], cell phone [7], X-ray medical machine [8], etc. One of important structure is P-N structure, the structure consists of N-type material and P-Type material [9]. The ion of both sides will be established across from N-type to P-type with positive bias. In this paper will study MSM structure by this structure use in photo detector area. MSM structure has 2-metal side by between metal is semiconductor material. [10]

Irradiation with high energy is one of process use for change semiconductor device characteristics. Soft radiation flash exposure (SRFE) is new technique use for increase semiconductor device performance by irradiate radiation on device few second by many times [11]. The overall dose of radiation on device may same with high energy exposed but soft radiation will not impact to device lattice and structure.

The goal of this paper in study impact of SRFE process on MSM photodetectors properties and investigate optimize process for MSM structure in the future. [12]

2. Experimental procedure

MSM device structure is Al/n-Si/Al and fabrication process step follow: (i) N-type (100) cleaning process with DI water and ultrasonic for 5 mins, (ii) Platinum-doped on back side of silicon, (iii) RTA annealing process for 5 mins at 800 °C, (iv) Al coating process on top of device, (v) Open active area by photolithography process, (vi) RTA process under 800 °C for 5 mins and (vii) check adhesion of metal contact by using 3M sticker then pull off. [13] MSM photodetector structure and SRFE process show in Figure. 1. Device after fabricate will measure electrical characteristics compare with after exposed by Roentgen irradiation for several times. The energy of radiation will using around 110 keV by exposed on top of device. Device will measured photocurrent in dark environment by sweep voltage from -10 to 10V, step 0.1V. [14]

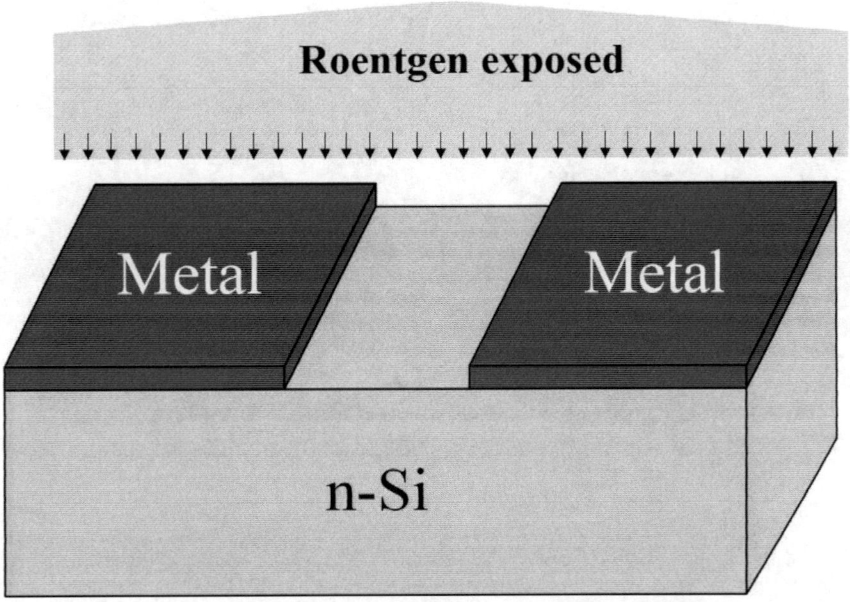

Figure 1. MSM photodetectors device structure.

3. Results and Discussions

Atom doping concentration by ion implantation process show in Figure 2. This process use high energy doped Pt atom on silicon substrate and may generate defect on surface of substrate. The concentration of doping atom show high density around 2-3 μm.

Figure 2. Simulation doping Pt atom by ion implantation process.

Figure 3. show Pt atom doped on back side of n-Si. Doping atom direction diffused to n-Si substrate by unpredict direction depend on concentration and lattice vacancy. Pt atom diffused deep into substrate around 2-3 μm and diffused to all area of substrate after RTA process [15]. Sheet resistance of n-Si substrate change after Pt-doped process that can confirm Pt atom diffuse to n-Si structure.

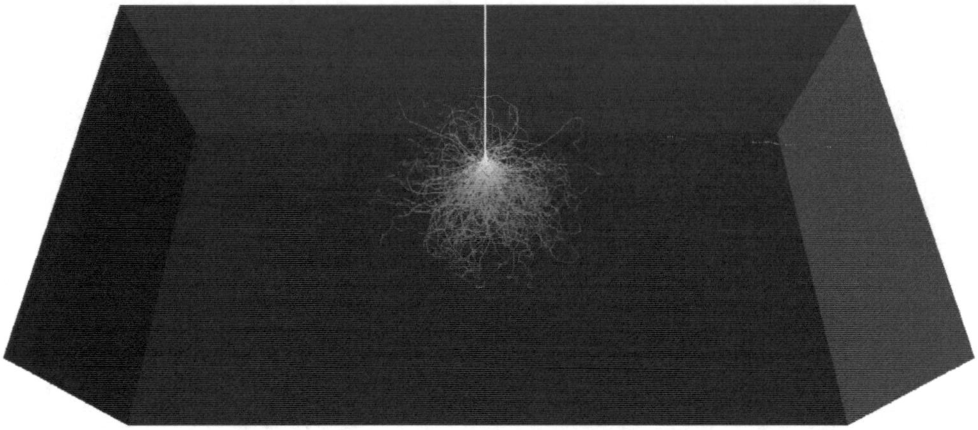

Figure 3. Platinum atom direction while doped in n-Si.

Capacitance-voltage characteristics (C-V) of MSM photodetectors before and after irradiation process show in Figure. 4. The results show capacitance various bias voltage from -10 to 10V, step 0.1V. After device irradiated by Roentgen show capacitance decrease but not obvious by capacitance decrease around 10 pF at 0V. The capacitance decrease may from radiation change device properties, radiation may impact to defects from Pt doping process and host defects [16]. The defects from fabrication process such as impurity, extended defects and point defects. One of the reasons to generate defects is ion implantation process, ion implantation using high energy for doping atom to substrate. Type of defect occur from ion implantation such as vacancy defects and interstitial defects. All defects can reduced device performance. [17]

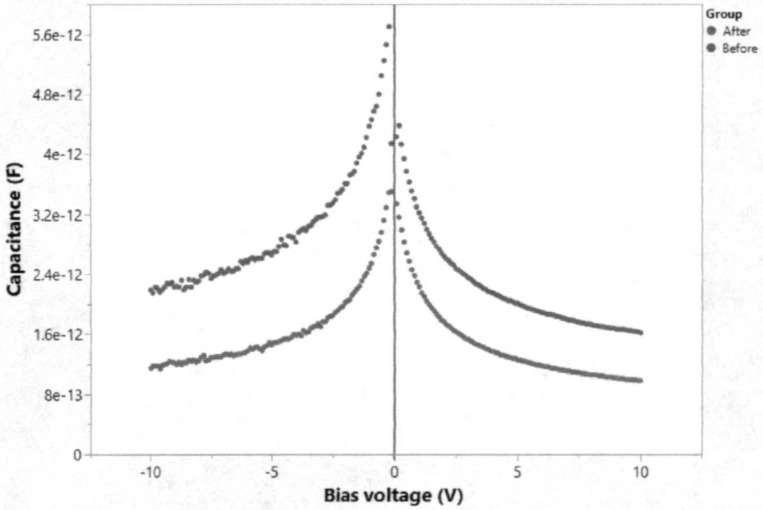

Figure 4. Capacitance-voltage characteristics of MSM device before and after irradiation by Roentgen.

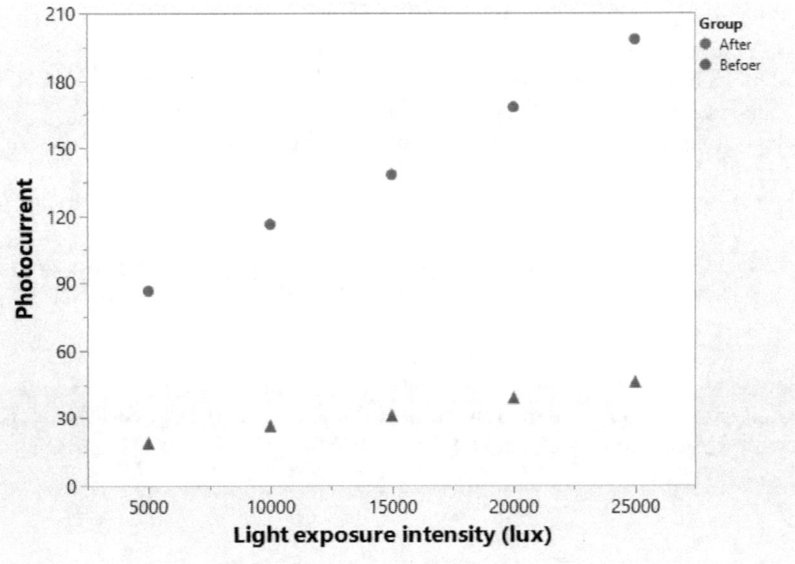

Figure 5. Photocurrent of MSM photodetector before and after exposed by Roentgen.

Figure 5. show photocurrent of MSM device by various intensity of light from 5000 to 25000 lux, step increase 5000. The results show after irradiation by Roentgen photocurrent show increase trend for all of intensity. Radiation may interact with device lattice and structure that help for device performance.

From photocurrent of MSM device results show that photocurrent before irradiate by Roentgen in range around 15-45 μA. Photocurrent increase around 4 times after irradiated, the increasing of photocurrent may relate with irradiation process. [18]

4. Conclusions

MSM photodetector device is one of design we use for many applications and would like increase performance. Capacitance-voltage characteristics of device after fabrication show around 5.6 pF and decreased after irradiate by Roentgen by show around 3.4 pF at bias voltage 0V. Photocurrent of device show increased around 4 times for all of light exposure intensity (lux). The increasing of photocurrent may help for improve performance of device; however, we have study more impact from irradiation process on MSM photodetector device. Although, photocurrent of MSM show significant increase and we would like to proof with other parameters to confirm affect from radiation such as carrier concentration, activation energy and Pt atom doping characteristics in n-Si substrate.

References

[1] David P K 1963 *Adv. Electron. Electron* vol 18 p 167-250
[2] Chris J M 1999 *Microelectron. Reliab.* Vol 39 p 731-739
[3] Riya T, Ashok R, Zhao Z J, and Yung K K 2021 *Mater Sci Semicond* vol 134 pp 106032
[4] Kemmer J and Lutz G 1988 *Nucl. Instrum. Methods Phys. Res. A: Accel. Spectrom. Detect. Assoc. Equip.* vol 273 pp 588-598
[5] Ying W, Changchun Z, Chunyu W and Junhua L 2005 *Microelectron Reliab* vol 45 p 3-4
[6] Bruno G 2017 *C. R. Phys.* vol 18 pp 453-461
[7] Jukka P V and Veli L 2005 *Solid State Electron.* vol 49 pp 1714-1721
[8] Ayzenshtat G I, Babichev E A, Baru S E, Groshev V R, Savinov G A, Tolbanov O P and Vorobiev A P 2003 *Nucl. Instrum. Methods Phys. Res. A: Accel. Spectrom. Detect. Assoc. Equip.* vol 509 pp 268-273
[9] Takashi S, Takashi K and Hideo I 2016 *Superlattices Microstruct* vol 99 p 165-168
[10] Munkhsaikhan Z, Zagarzusem K, Shim H Y, Jongkhan W, Sung N L and Chel J C 2017 *Microelectron Reliab* vol 69 p 60-65

[11] Narong S, Itsara S and Surasak N 2021 *IOP Conf. Ser.: Mater. Sci. Eng.* Vol 1070 p 1-4
[12] Rozana H, Yixuan C and Yi L 2013 *Appl. Phys. Lett.* Vol 102
[13] Narong S, Itsara S and Preecha 2016 P Y *Far East J. Electron. Commun.* Vol 16 p 451-458
[14] Narong S, Itsara S and Surasak N 2018 *Suranaree J. Sci. Technol.* Vol 28 p 1-4
[15] Hanfeng Z, Caichi L, Qigang Z, Jing W, Qiuyan H and Hongdi Z 2004 The Fourth International Workshop on Junction Technology, 2004. IWJT '04
[16] Songul D, Fikriye S K, Hulya D, Guven T and Yilmaz S 2022 Radiat. Phys. Chem. Vol 193 p 1099992
[17] Shuaiwei F, Lei Y, Zhonglong W and Guoying G 2020 *Comput. Mater. Sci.* Vol 173
[18] Michael A H, Irina V, Ian G H, Alasdair S 2020 Org. Electron. Vol 85 p 105890

Acknowledgments
The authors would like to thanks Suan Sunandha Rajabhat University (SSRU) and King Mongkut's Institute of Technology Ladkrabang (KMITL).

Journal of Physics: Conference Series **2331** (2022) 012009 doi:10.1088/1742-6596/2331/1/012009

Research on Hardware System Design Method of the Intelligent Vehicle Based on Dual Sensors

Wenliang Li

College of Mechanical and Electrical Engineering Wuhan University of Technology, Wuhan, Hubei Province, China

wenlianglimail@foxmail.com

Abstract. With the rapid development of modern society, the application of intelligent unmanned vehicles is becoming more and more popular. Based on the scene of unmanned vehicle patrol operation in an intelligent warehouse and simplifying it, this paper designs an image processing and perception of driving road conditions with ATM89C51 single-chip microcomputer as the core, camera, and infrared sensors to realize the detection of car trajectory. Two groups of four DC motors are set to drive, and the turning is carried out through the differential speed of the wheel motors on both sides. The turning radius and turning time are shortened through the independent control of the motors on both sides. The L298N full-bridge motor drive core is used to control the speed of the motor, to realize the accurate control of the trolley movement, and lay the foundation for the intelligent operation of the trolley in more scenarios. The accuracy of car navigation and motion control is improved through dual sensors.

1. Introduction

With the rapid development of modern society, the research on the intelligent vehicle is also deepening. In 1939, General Motors Company first put forward the concept of unmanned vehicles, and Carnegie Mellon University developed the world's first truly driverless vehicle in 1984 [1]. China's unmanned car started late, and the initial stage was mainly concentrated in colleges and universities. After a period of development, some major enterprises also announced to enter the unmanned car industry, and the domestic unmanned car industry has developed rapidly. As a reduction system of intelligent vehicles, the intelligent vehicle system has become a good experimental platform in the process of intelligent vehicle research and development, which has the characteristics of low cost and high simulation[2]. The research of intelligent vehicle systems has become one of the hot research directions in many fields [3]. During the development of the intelligent vehicle, the requirements for the control accuracy and road information processing ability of the car are continuously improved. Liu Peng et al. [3] controlled the motor and steering gear through MK60FX512VLQ15 microcontroller to realize steering; Ye Mengjun et al. [4] used SD5 digital steering gear to directly connect to the pin of single chip microcomputer, and adjusted the duty cycle of steering gear through the control signal of single chip microcomputer to realize the steering function of steering gear; Gao Hong et al. [5] improved the response speed of steering by lengthening the steering arm with aluminum alloy sheets; Sheng Liuqing [6] improved Hebb learning rules to make the single neuron adaptive PID control algorithm have stronger anti-interference and better stability; Yan Ke et al. [7] proposed a pre judgment method for open circuit elements and a switching scheme between electromagnetic and camera in open circuit, which improved the accuracy and real-time of path image processing of intelligent vehicle in complex environment; Huang Dingjian [8] and others completed the design of an intelligent car for obstacle detection and ranging based on vision and ultrasound by fusing ultrasonic sensors and visual images, and realized the car's active obstacle avoidance and lane change function through relevant experiments; Yang Pengshuai [9] and others used gray-scale sensor to reduce the interference of external environment light source on camera image processing, and used cascade PID closed-loop to control Mecanum wheel to realize rapid and stable operation of the car. In the aspect of intelligent vehicle target detection algorithm, Y. Wang, D. Chen et al.[10] proposed an algorithm based on Hough transform: the length of the edge detection line segment is counted, and then the missing line points of the lane line are identified, and the maximum likelihood estimation is used. Measure the difference between the actual lane line and the model lane line, and then use the maximum likelihood method to detect the lane line. The AlexNet algorithm developed by the Canadian team[11] is improved based on neural networks, which is faster than traditional image processing

This paper mainly discusses the system modeling, hardware circuit principle and implementation mode, software execution process architecture, and program implementation. The first part briefly introduces the kinematics and dynamics model of the vehicle system, and mainly describes the principle and modeling of the intelligent car system; In the second part, by putting forward the ground track detection method of double sensors and the control strategy of differential speed adjusting motor speed based on PID control, the accuracy of car line patrol operation is improved, the turning radius and time of car are reduced, the response speed of car operation is accelerated, and the control function of the intelligent vehicle traveling along the planned path and real-time attitude adjustment is completed; The third part expounds the overall operation control logic of the car through the implementation of relevant software programs, and obtains the simulation results through the simulation of relevant platforms.

2. Principle and modeling of intelligent vehicle

2.1. Kinematics model of intelligent vehicle

The intelligent vehicle adopts the differential drive model, and the dynamic model of the car is obtained by combining the rotation mathematical model of the motor with the position coordinates of the car.

Establish a kinematic coordinate system for the trolley, abstract it into a simple four-wheel model, and set its centroid coordinates as (x, y), as shown in Figure 1:

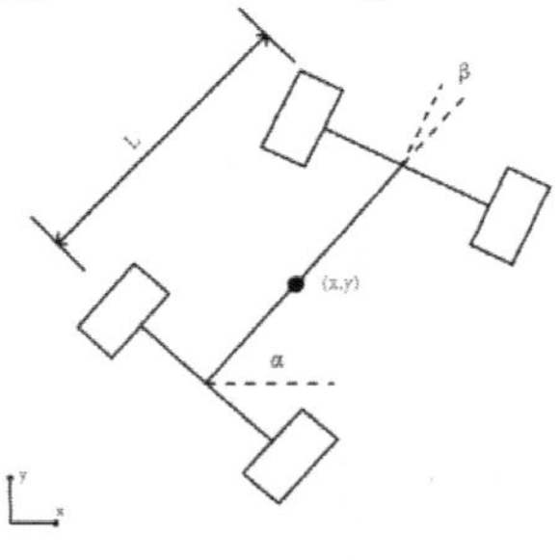

Figure 1. Schematic diagram of trolley kinematics model.

The kinematic model of the trolley is established according to the above figure, as shown in equation 1:

$$\begin{cases} \dot{x} = v\cos\alpha \\ \dot{y} = v\sin\alpha \\ \dot{\alpha} = v / L\tan\beta \end{cases} \tag{1}$$

The dynamic model of the trolley can be obtained by combining the kinematic model of the trolley with the motor motion equation (see equation 2), as shown in equations 2 and 3:

$$T = J\frac{d\dot{\theta}}{dt} + T_f \tag{2}$$

$$\begin{cases} T = J\dfrac{d(x/r\cos\alpha)}{dt} + T_f \\ T = J\dfrac{d(y/r\sin\alpha)}{dt} + T_f \\ T = J\dfrac{d(\dot{\alpha}L/\tan\beta)}{dt} + T_f \end{cases} \tag{3}$$

Where T is the motor torque, T_f load torque, the J-the moment of inertia of the motor rotor and load, (x, y) - the geodetic coordinate of the center of mass of the car, r-radius of the car wheel, α— azimuth, β— front-wheel angle.

The motion control of the intelligent vehicle is realized by the control of the driving motor and servo actuator. In an ideal situation, it can be considered that four tires roll around the same center of a circle, and the rigid body dynamics model is established based on the above theory. Then the running angle of the trolley steering gear is twice the actual rotation angle of the trolley center of mass, and the discrete motion model formula is:

$$\begin{cases} x(t+1) = x(t) + v(t)\times\Delta t\times\cos\left[\alpha(t)+\omega(t)\times\Delta t/2\right] \\ y(t+1) = y(t) + v(t)\times\Delta t\times\sin\left[\alpha(t)+\omega(t)\times\Delta t/2\right] \\ \alpha(t+1) = \alpha(t) + \omega(t)\times\Delta t \end{cases} \tag{4}$$

Where V is the linear speed of the wheel, ω Is the steering angular speed of the steering gear, X (T) and Y (T) are the coordinates of the trolley, α (t) For the steering angle of the smart car, Δ T is the sampling period.

According to the above formula, the intelligent vehicle system plans the traveling path of the trolley according to the track position of the actual ground route and adjusts the movement state of the trolley by changing the speed of the motor, to achieve the purpose of moving the trolley along the track and realize the accurate control of the movement of the trolley.

2.2. Motor control model

The intelligent vehicle system is driven by a DC motor. The control of the intelligent vehicle system is essential to control the speed and torque of the driving motor. For DC motor, the relationship between torque and speed follows the basic motor motion equation, as shown in equation 2. Therefore, based on the control requirements of motor torque and speed, the relationship between speed and voltage is obtained after comprehensively considering torque and speed, torque and current, back EMF and speed, and static characteristics of the motor, as shown in equation 5:

$$\dot{\theta} = u_a / K_e - R/(K_t K_e)\times T_f \tag{5}$$

Wherein, UA armature voltage, R is the resistance of the driving circuit, K_e is the back EMF coefficient, K_t is the torque coefficient, and T_f is the load torque.

When the intelligent vehicle reaches the constant speed state after the motor is started, the mechanical and control characteristics of the driving motor show regular changes, as shown in Figure 2. In order to clarify the speed regulation process of the trolley drive motor in detail, speed regulation auxiliary lines are added in the figure. According to the motor operation control characteristics, at the same speed, when the load increases, the armature voltage increases; It can be seen from the mechanical characteristics that there is a positive correlation between speed and voltage under the same load conditions. Therefore, in the motion control of the trolley, when the load is certain, the motor speed is controlled by changing the input voltage of the motor, and then the motion attitude of the intelligent vehicle is controlled.

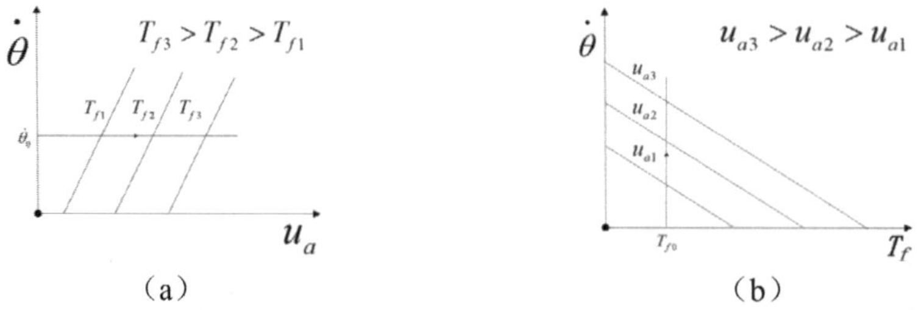

Figure 2. Working characteristic curve of the motor.

The classical PID control algorithm is adopted for the motor, and the transfer function block diagram is designed in combination with the dynamic characteristics of the motor, as shown in Figure 3. When the duty ratio of the motor receiving the PWM signal is large, the motor inputs voltage increases, the motor torque increases, and the trolley accelerates. When the duty ratio of the PWM signal decreases, the motor input voltage decreases, the torque decreases, and the trolley decelerates. The control relationship is shown in Figure 4. Take the torque as the horizontal axis and the rotation speed as the vertical axis to establish a four-quadrant control diagram. When the trolley is faced with road conditions such as curves, the motors on both sides of the trolley receive PWM signals with different duty cycles, the speed of the motors on both sides will change, to control the turning of the trolley and adjust the real-time position of the trolley.

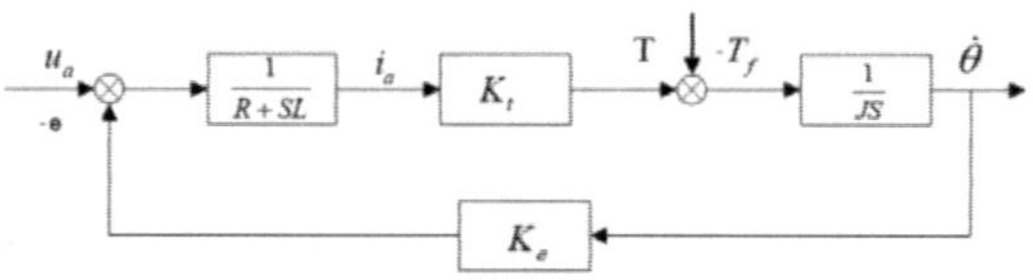

Figure 3. Block diagram of the control transfer function of driving motor.

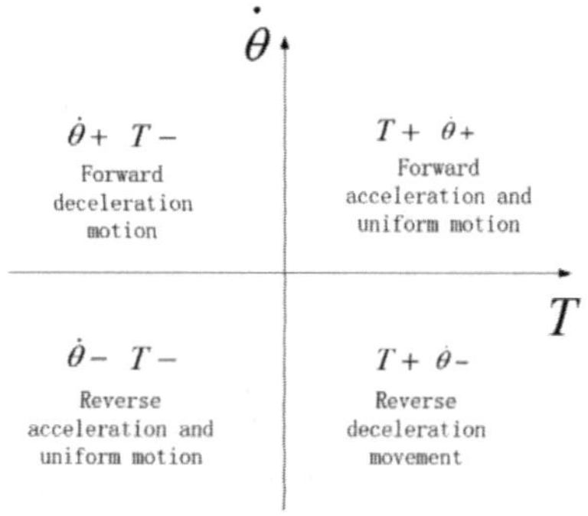

Figure 4. Four-quadrant control diagram of motor.

3. Overall hardware system design of the intelligent vehicle

The preset driving road of the intelligent vehicle is a road with a width of 45cm. The edge and center of the road are provided with a boundary line with a width of 2.5cm. The preset driving road is provided with a curve with a radius of 55cm and a ramp with an inclination of 10 °. When the intelligent vehicle is running, it detects the centerline of the road and patrols the line. The requirements for relevant operating conditions are shown in Table 1:

Table 1. Relevant design requirements of trolley.

Requirement of designing	Parameters
Maximum current and rated voltage of the motor	**3A、12V**
The maximum turning angle of the vehicle	**15°**
Trolley climbing angle	**-20°~20°**
Camera frame rate, pixel	**752×480、60fps**
MCU bandwidth	**12MHz**

3.1. Overall system design

The hardware system mainly includes the controller with ATM89C51 single-chip microcomputer as the core, 7.2V power supply module, input device, MT9V032 camera with reflective infrared sensors as the image acquisition module with reflective infrared sensors, photoelectric code disk speed detection module, 12V DC motor drive module, communication serial port, and external display equipment. According to the functions of the above modules, the hardware part can be divided into the power module, information acquisition part, information processing part, real-time control part, and input/output part. The above parts together constitute a closed-loop control system, as shown in Figure 5.

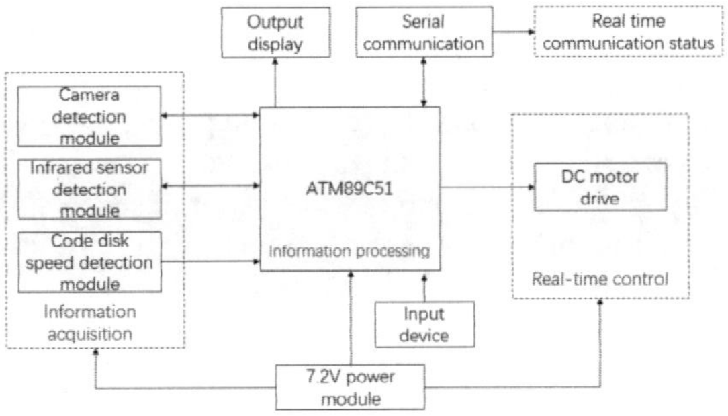

Figure 5. Hardware system diagram of intelligent vehicle.

The intelligent vehicle information acquisition part obtains the environment of the intelligent vehicle and identifies the ground track line in real-time through the camera, supplemented by infrared sensors to detect the road centerline in real-time, and uploads the detection results to the single-chip microcomputer through the RS232 interface. Combined with the photoelectric code disk, the motor speed is sampled, and the motor speed is converted into the traveling speed of the intelligent vehicle through the kinematic relationship. After AD conversion, the relevant information is uploaded to the single-chip microcomputer, and then the single-chip microcomputer processes the relevant information in real-time and sends the instruction to the driving motor to control the motor speed, to ensure that the trolley moves forward along the ground track line and realize the control of the trolley movement.

To better achieve the purpose of rapid response, the trolley is designed to use a four-wheel drive. By adjusting the speed of wheels on both sides, the movement objectives of the trolley such as start, stop, and steering is achieved. At the same time, the differential is installed on the front wheel to ensure the normal steering of the trolley. It is controlled by a software program and combined with hardware architecture, to realize the function of automatic tracking on the solid line and driving along the line.

3.2. Hardware design of control system based on a single-chip micro-computer

The hardware circuit of the intelligent vehicle selects ATM89C51 single-chip microcomputer as the controller, supplemented by a clock circuit, power circuit, reset circuit, and BDM interface to form the minimum system circuit of the single-chip microcomputer. Among them, the clock circuit adopts a 12Mhz quartz crystal oscillator, the power supply voltage is 5V, and the BDM interface is used for users to download and debug programs to the MCU. The main controller processes the input signals of the camera, infrared sensors, and photoelectric code disk, processes the signals according to the program downloaded by the user, and then sends them to the motor control chip L298N. The motor chip outputs a PWM voltage signal to control the operation of two groups of driving motors, to control the movement of the small car.

3.3. Acquisition module for road information

The road information of the intelligent vehicle is obtained mainly through the MT9V032 camera to prejudge and identify the track line of the road ahead, supplemented by the TCRT5000 reflective infrared sensor for real-time position detection, which can accurately drive along the track line on the premise of ensuring the forward-looking movement of the car, effectively reducing the tracking deviation caused by single-camera road recognition and reducing the possibility of car yaw.

MT9V032 camera resolution is 752 × 480, the frame rate is 60fps, and the PAL signal is output. To improve the CPU performance, the software binarization method is directly used to process the collected video image information. LM1881 is used as the synchronous separation chip of video signal. When installing the camera, rotate it by 90° to obtain a large forward-looking perspective.

TCRT5000 reflective infrared sensor is mainly composed of a light-emitting diode and photosensitive triode. When in use, the light-emitting diode emits infrared. When the light signal at the transmitting end irradiates the black track line, the photosensitive triode is interrupted due to the weak intensity of the reflected light signal, and the module outputs a low level. When the light signal irradiates to the position outside the track line, the reflected light signal is strong, and the triode is turned on and outputs a high level [12].

According to the above laws of the infrared sensor, the real-time detection module of the infrared sensor is designed. The module is mainly used to reduce the probability of deviation from the track line during the movement of the car and ensure that the car runs along the line. Considering the possible yaw problem of the trolley during operation, a group of three infrared sensors, which is 6mm away from the ground, is set and fixed at the front of the trolley, as shown in Figure 6. When the trolley runs along the middle of the road, the middle sensor detects the road track line and returns the low-level signal, and the sensors on both sides return the high-level signal. At this time, the motor speed on both sides of the trolley remains the same, and the trolley runs along the track line; When the position deviation occurs during the movement of the trolley, the sensor on one side returns to the low level and the other sensors return to the high level. At this time, there is a deviation in the attitude of the trolley. Therefore, by increasing the motor input voltage on the opposite side of the low-level sensor, the motor speed is increased, and then the wheel speed of the trolley is adjusted to adjust the attitude of the trolley so that the trolley can travel accurately along the line and reduce the possibility of yaw of the trolley.

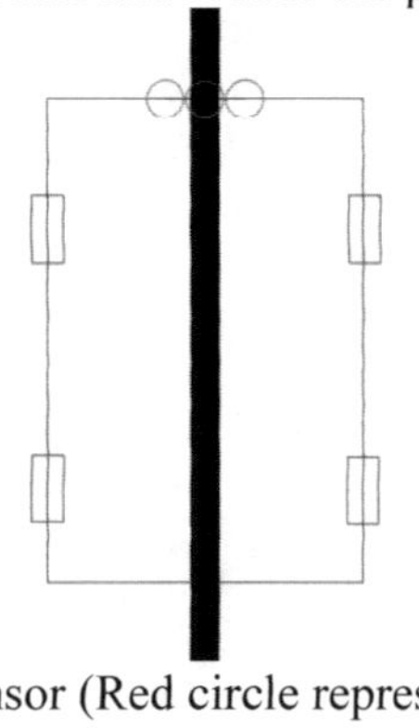

Figure 6. Layout diagram of infrared sensor (Red circle represents infrared sensor and black solid line represents ground track)

3.4. Motor drive module

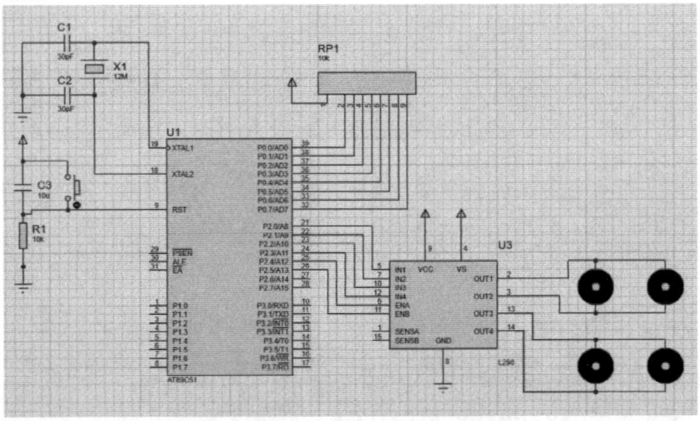

Figure 7. Motor drive diagram

The intelligent vehicle adopts a 12V DC motor as the drive. Compared with the stepping motor, the software control of the DC motor is simpler and easier to control the motor. The motor control circuit adopts an H-bridge circuit to adjust the motor speed through digital adjustment.

H-bridge circuit has the characteristics of good speed regulation characteristics, wide speed regulation range, strong overload capacity, and can withstand frequent load impact. It can also realize frequently rapid start, braking, and reverse rotation, which is suitable for the motion scene of the trolley. By adjusting the duty cycle of the gate level PWM of the driving tube, the digital regulation mode directly adjusts the on and off time of the Darlington tube, and indirectly adjusts the voltage applied to the motor to control the motor speed. In this design, the L298N full-bridge motor drive core is selected to be connected with the single-chip microcomputer, and the motors on the same side of the trolley are connected with the drive core after being connected in parallel, that is, the L298N motor drive core drives the four motors on the left and right sides of the trolley at the same time, as shown in Figure 7. This speed regulation method has higher control accuracy, which is easier to realize the fine-tuning of the trolley position and is more convenient to realize the overall automatic control link, and lays a foundation for the intelligent improvement of the trolley in the future.

3.5. Detection module for vehicle speed

The photoelectric code disc is used for speed measurement. The photoelectric code disc can output two groups of pulses with a phase difference of 90 degrees between a and B by using the rotary encoder with two-way output. Since the phase difference between A and B is 90°, the forward and reverse rotation of the encoder can be distinguished by comparing the phase positions of A and B, and the motor speed and rotation direction can be measured by two groups of pulses. When the trolley is moving, the photoelectric code disk of the motor continuously generates the corresponding pulse signal. After frequency voltage amplification of the sampled pulse signal, the voltage signal proportional to the rotating speed of the rotating shaft is obtained to measure the running speed of the trolley.

3.6 Power module

The intelligent vehicle is powered by a 7.2V, 3A·h rechargeable battery. Because different components of the smart car require different voltages, the power chip is used to boost or depressurize the output voltage accordingly. The 7.2V power supply provides the input voltage source for the DC-DC power chip, and then the power chip set 3.3V, 5V, and 12V voltage respectively to control the voltage of each part. Power supplies for MCU, infrared sensor, photoelectric code disk, and external equipment by configuring 5V voltage; power the MT9V032 camera by configuring 3.3V voltage; the 12V voltage is configured to supply power to the DC drive motor, and the specific voltage distribution is shown in Figure 8.

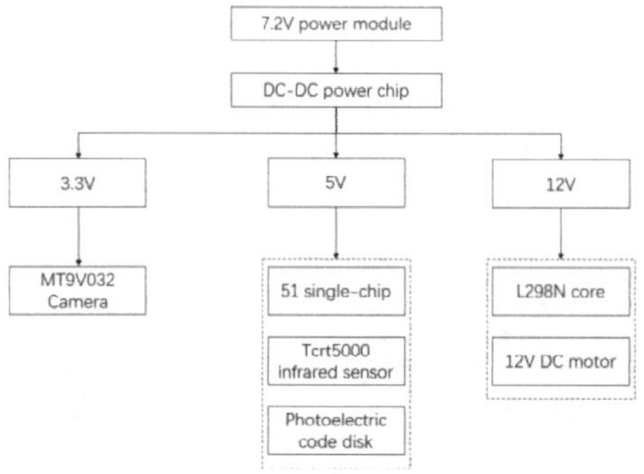

Figure 8. Power supply voltage distribution diagram

4. Software design procedure of intelligent vehicle system

4.1. Software design of trolley line inspection module

The key to an intelligent vehicle system to realizing line patrol operation is the rationality of the control algorithm. A reasonable control algorithm can effectively enhance the stability of an intelligent vehicle system and improve the operation efficiency of the system. The main function of the intelligent vehicle system is line patrol operation, that is, independently track the ground track line and run to the designated position along the line. The overall flow chart of the system software is shown in Figure 9.

The system road tracting is mainly completed by the camera module. Due to the single situation of the ground route and considering the limited processing speed of the single-chip microcomputer, the method of interlaced sampling is adopted when the camera collects site information. Considering the forward-looking problem of intelligent vehicle operation, the strategy of far more and near less interlaced sampling is adopted to maximize the computing power of the single-chip microcomputer [13]. In the camera information processing, the collected image is gray, and a threshold is required to determine the position of the black line. Due to the possible impact of the external environment such as reflection on the image acquisition, the dynamic threshold method is used to process the image information in the image processing [14]. After processing the discrete image information collected by the camera, the single-chip microcomputer judges the road information and then outputs the relevant discrete control signal to the driving motor module. Facing the road curve, the single-chip microcomputer calculates the curvature of the road centerline according to the image information, adjusts the control parameters of the motor, and realizes autonomous turning.

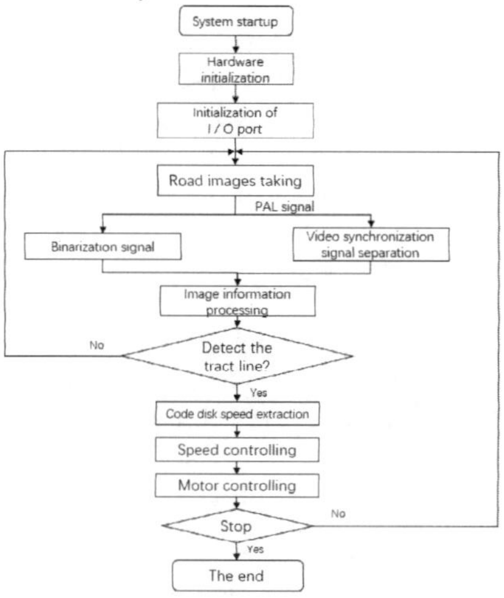

Figure 9. Overall flow chart of the software system

4.2. Software design of real-time detection module for road information

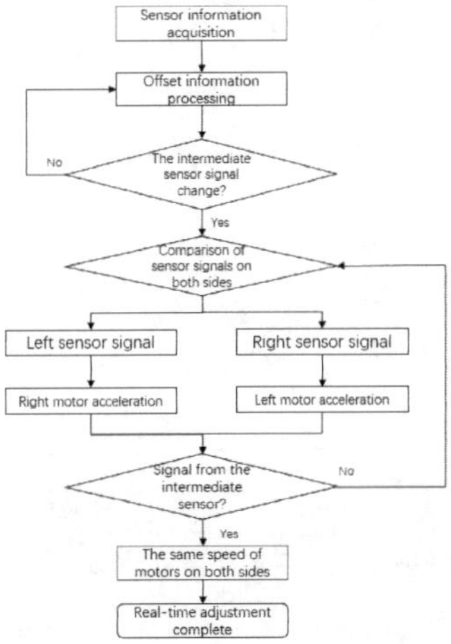

Figure 10. Flow chart of the real-time detection algorithm

In the process of an intelligent vehicle system running algorithm, a single camera tracking process is difficult to realize that the car travels along the track line accurately. When the error accumulation caused by camera tracking, the car is easy to deviates from the track. Therefore, based on the tracking scheme of the monocular camera, this paper proposes a scheme of using three infrared sensors to obtain the real-time position of the road track line to assist the car to patrol the line to improve the tracking accuracy. As shown in Figure 10 the software design procedure of the road real-time detection module is when the car is driving in the middle of the road, the middle sensor outputs a low level with both sides output a high level, and the motor speed on both sides of the car remains the same; When the trolley leans left or right to the track line, the right or left sensor detects that the track line is at a low level. At this time, the level signal is output to the single-chip microcomputer. After the single-chip microcomputer judges the yaw condition in combination with the level signal of the middle sensor, it sends corresponding instructions to the driving motor to control the motor speed on both sides. When both the single side sensor and the middle sensor are at a low-level signal, the trolley yaw angle is small, fine-tune the motor speed on the opposite side to correct it; When only one side sensor is in the low-level signal, the yaw angle of the trolley is large, quickly adjust the opposite side motor speed for attitude correction. That is, when the left (right) sensor outputs a low level, increases the motor speed on the right (left) side of the trolley, quickly adjusts the real-time position of the trolley, and ensures that the trolley is generally traveling along the road centerline. When the middle sensor returns to a low level and the sensors on both sides are at a high level, adjust the rotation speed on both sides of the trolley to be consistent, to realize the real-time position and attitude correction of the trolley, and lay a foundation for the subsequent application of other scenes of the trolley.

5. Modeling and simulation

The intelligent vehicle performance in this paper is tested and compared based on the MATLAB simulation platform and offline actual operation. The results are shown in Figure 11. After the actual calculation and verification, the maximum error between the simulated trajectory and the measured trajectory, and the preset trajectory is about 6%. Therefore, the system has high accuracy, so this design has high feasibility in the accurate line finding driving of the intelligent vehicle.

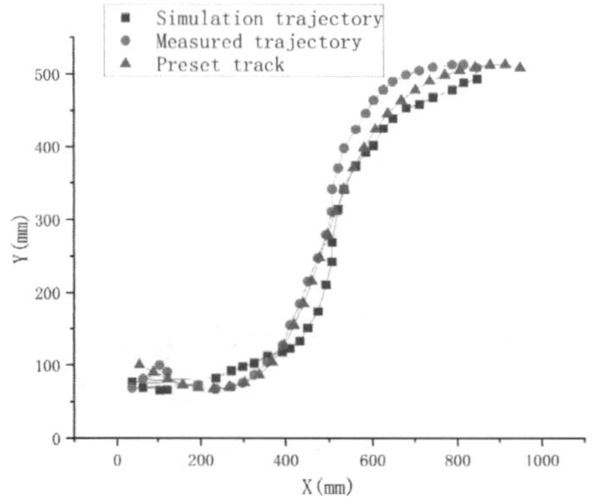

Figure 11. Comparison between simulation test results and measured results and preset trolley trajectory

6. Conclusion

In this paper, a hardware system design method for the intelligent vehicle with two sensors is proposed, it adopts a DC motor as the executive element, MT9V032 camera, and TCRT5000 infrared sensor as the detection element, and ATM89C51 single-chip microcomputer as the main control chip. The hardware and software design of intelligent vehicle is completed, thus, the high-precision autonomous line patrol and real-time attitude adjustment of the intelligent vehicle are successfully realized, it also has certain practical value in the intelligent management and material handling of the unmanned warehouse.

Reference

[1] Yu Daoyang 2021 Design of four-wheeled trolley control system based on visual line inspection in the "NXP" Cup Master's Thesis, Nanchang University
[2] Sheng Liuqing 2021 Application of adaptive PID control algorithm in intelligent vehicle control system. Journal of Beijing Institute of Graphic Design 05 147-150
[3] Liu Peng, Sun Defan, Wang Tao & Jia Yulei 2021 Design of smart car based on photoelectric sensor target seeking Electronic Design Engineering 21 8-11
[4] Ye Mengjun, Huang Yanglai and Lei Gaihui 2020 Design and implementation of an intelligent vehicle system based on OV7725 camera Journal of Hubei Normal University (Natural Science Edition) 02 82-87
[5] Gao Hong, Wang Yuchen and Li Huiying 2019 Camera-based mechanical modification and optimization of smart cars Mechanical Design S2 196-199
[6] Sheng Liuqing 2021 Application of adaptive PID control algorithm in intelligent vehicle control system Journal of Beijing Institute of Graphic Design 05 147-150
[7] Yan Ke, Wang Qi, Wang Weixi, Shen Tong and Li Yi 2021 Trace-based image edge extraction and circuit break recognition for smart cars Automation and Instrumentation 04 12-16+20
[8] Huang Dingjian, Yang Huashan, Lin Chenyao, Zhong Yong and Wang Shangyu 2021 Obstacle Avoidance Design of Intelligent Car Integrating Ultrasonic Sensing and Visual Image Journal of Fujian Institute of Technology 06 511-514+555
[9] Yang Pengshuai, Wang Qi, Wang Weixi, Zhou Jingyu and Gao Jinke 2022 Research on the control strategy of intelligent vehicle tracking based on Mecanum wheel Industrial Control Computer 02 112-115
[10] Yue Wang, Dinggang Shen and Eam Khwang Teoh 2000 Lane detection using spline model Pattern Recognition Letters 8
[11] Jia, D. , Wei, D. , Socher, R. , Li, L. J. , Kai, L. and Li, F. F. 2009 ImageNet: A large-scale hierarchical image database pp.248-255
[12] Yan Junxu and Hou Chao 2012 Intelligent tracking car based on infrared reflective photoelectric

sensor Shanxi Electronic Technology 02 34-35+47
[13] Huang Zhiqiang 2011 Design of segmented PID control algorithm for intelligent vehicle based on CCD camera. Electronic Design Engineering 02 55-57
[14] [Yuan Yadong 2009 Development of CCD-based smart car Master's thesis, Hefei University of Technology

A BCF Bionic Robot Fish Driven by A Dielectric Elastomer Actuator

Zhi-Wei Jiao, Hao-Yu Wang, Bin Luo, Wei-Min Yang and Yuan Yu*

College of Mechanical and Electrical Engineering, Beijing University of Chemical Technology, Beijing, 100029, China

The corresponding author's e-mail address: yuyuanjd@263.net

Abstract. Bionic robot fish have been widely used in various underwater exploration activities due to their advantages of low noise, low resistance, high flexibility and high concealment. A dielectric elastomer (DE) is a soft material that significantly deforms under the action of an electric field. It has the characteristics of large strain, flexibility, high energy density and fast response speed, which lead to its unique advantages in the actuation of bionic fish. In this paper, the actuation characteristics of a dielectric elastomer are used to design and prepare a limit-type antagonistic dielectric elastomer actuator (DEA). A BCF bionic robot fish is proposed based on the DEA, which has a maximum swimming speed of 22.7 mm/s.

1. Introduction

With the increasing complexity of exploration environments, the limitations of traditional rigid robots will greatly reduce their environmental compatibility. Compared with traditional rigid robots, soft robots have unique advantages in complex exploration environments, such as a high degree of freedom, strong adaptability, and the ability to withstand large deformations [1]. In ocean exploration, soft robots can imitate the morphological characteristics of fish and have the characteristics of high mobility and flexibility, which lead to their wide application prospects.

To realize the application of soft robotics in bionic robot fish, researchers have developed various actuation methods for soft bionic robot fish. There are mainly four common actuation forms of bionic fish, including motor actuation [3], air/fluid actuation [7], Shape Memory Alloy actuation [8] and Ionic Polymer–Metal Composite actuation [10]. Zhang et al. developed a bioinspired bionic fish that uses flexible matrix composite (FMC) actuators for fin and body actuation [13]. C. Rossi et al. used shape memory alloy actuators (SMAs) to drive bionic fish [9]. Li et al. developed a bionic robot fish with a swinging pectoral fin using a dielectric elastomer actuator (DEA) [14]. Lou et al. used piezoelectric fiber actuators to develop a carp-like tail fin small propeller [16]. Andrew D. Marchese et al. designed and manufactured a pneumatic bionic robot fish [7], whose actuation is mainly based on multiple small air passages, and the bending and swinging of the fish tail is controlled by air pressure to realize underwater movement. S. Shatara et al. proposed a bionic fish driven by an ionic polymer-metal composite (IPMC) actuator [11].

Among the abovementioned actuator materials, the dielectric elastomer is a soft material that can produce significant deformation under the action of an electric field, and it is also called an artificial muscle. Due to the characteristics of large strain, high energy density, fast response speed, and light weight [17], it has high application prospects in the actuation of bionic robot fish. Based on DEAs, researchers have developed bionic fish with different motion patterns, such as imitation breaststroke robots [19], jellyfish robots [20] and pectoral fin swing robot fish [14]. To achieve faster and more efficient underwater motion, researchers have analyzed various types of fish propulsion models. Among them, BCF (body/caudal fin) propulsion is currently the fastest motion pattern of fish propulsion [21].

This study revealed the morphological characteristics of BCF fish in nature. By imitating the alternating contraction of muscles when fish swim, combined with the actuation characteristics of a DE similar to artificial muscles, a limit-type antagonistic dielectric elastomer actuator is proposed. The preparation method of the DEA is detailed, and an underwater test platform is built to test the swing performance of the DEA passive tail fin. CFD simulation of the DEA passive tail fin is performed using moving grid technology in the FLUENT environment, and the propulsion is calculated. The DEA is applied to the actuation of BCF bionic robot fish.

2. Design of the BCF bionic robot fish

2.1 Prototype model

The model of the robot fish mainly imitates the Carangidae BCF fish. The swing amplitude of the front part of the body is very small, and the passive caudal fin is driven by the swing of the back 1/3 of the body when swimming. The body part of the model and DEA were made separately and then assembled. Figure 1 shows a three-dimensional model of the robot fish after assembly. The body of the robot fish is in the shape of an inverted triangle and was formed by 3D printing. The lower part of the robot fish is filled with a solid material, and the upper part is hollow, similar to the principle of a tumbler, which allows the robot fish to maintain balance in the water. The maximum spatial size of the robot fish is 175 mm×50 mm×70 mm.

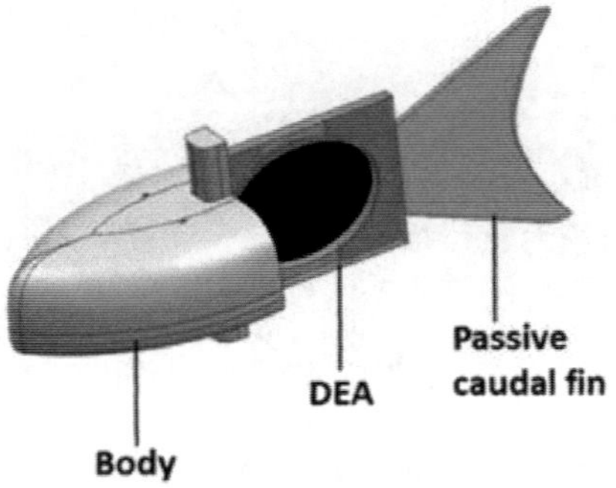

Figure 1. Three-dimensional model of the robot fish

2.2 Working principle of the DEA

The DE is composed of an elastomer film sandwiched between two flexible electrodes, which thus form a capacitor. The working principle is shown in Figure 2. When an electric field is applied, under the action of the Maxwell force, the DE film becomes thinner in the thickness direction but expands the area of the plane, and consequently, electrical energy is converted into mechanical energy.

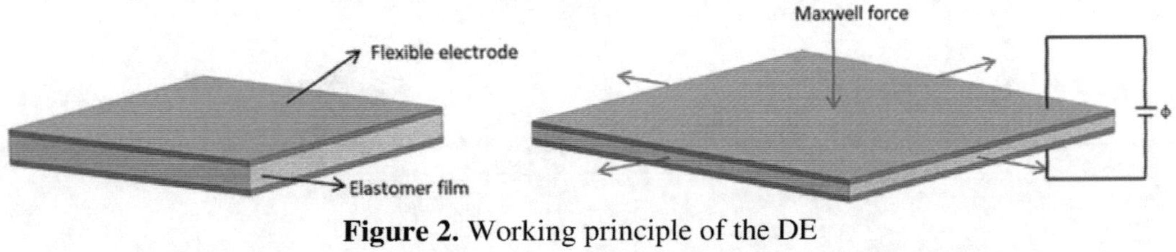

Figure 2. Working principle of the DE

The structure of the DEA is shown in Figure 3. Its main body is a skeleton composed of silicone, and a limit piece is placed inside the skeleton, which is similar to the spine skeleton of a fish body. There are two DEs on both sides of the skeleton, which are biaxially pre-stretched. The limit piece is used to limit the deformation of the skeleton in the Y-axis direction, thereby limiting the deformation of the DEs in the Y-axis direction. In this way, the bending deformation of the DEA is only reflected in the X-axis direction, and the contraction force of the DEs can be concentrated on the X-axis.

Figure 4(b) shows the top view of the Y-axis of the DEA when both sides are not activated. At this time, the contraction forces of the DEs on both sides are equal. When the right DE is activated and the left DE is not activated (Figure 4(a)), the expansion of the activated DE causes the contraction force of the right DE to decrease, while the contraction force of the left DE is not altered due to inactivation. Due to the unbalanced contraction force of the DEs on both sides, the DE with the larger contraction force on the left will actuate the DEA to bend to the left. Similarly, when the left DE is activated and the right DE is not activated (Figure 4(c)), the DEA bends to the right.

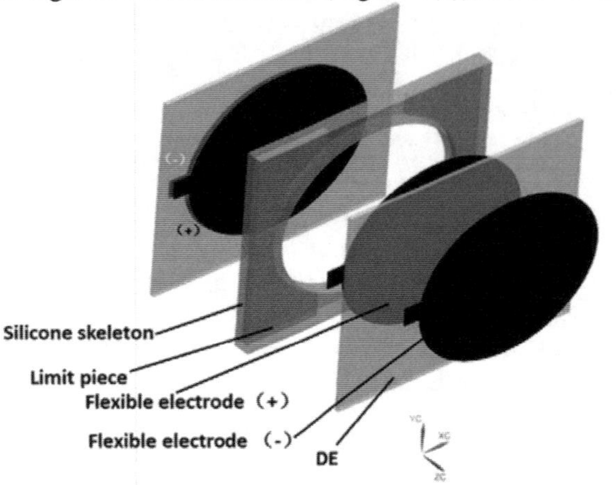

Figure 3. Structure of the DEA

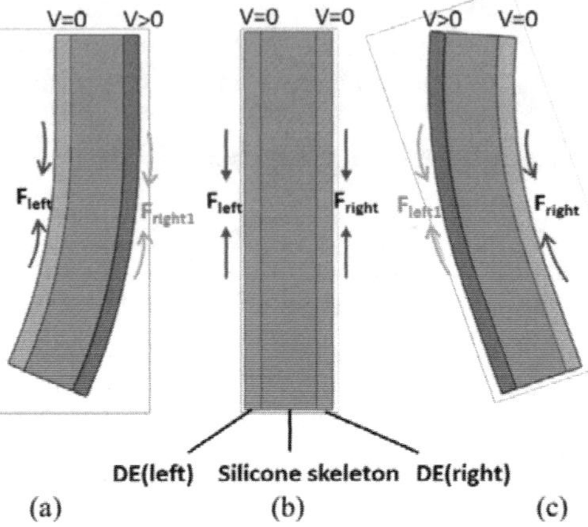

Figure 4. Working principle of the DEA

3. Materials and Methods

3.1 Fabrication process of the DEA

The fabrication process of the DEA is shown in Figure 5. The silicone skeleton is prepared by pouring. 3D printing technology is used to prepare the mold of the silicone skeleton and the limit piece. First, silicone (Dragon Skin 30) is poured into the mold, and the limit piece is placed in the silicone. After being kept in an oven for 40 minutes (45°C), the silicone skeleton is obtained by demolding. Then, the DE material (VHB4910, 3M Company) with a thickness of 1 mm is biaxially stretched to 4×4 times and fixed on the circular fixed frame. The upper and lower surfaces of the DE are covered with a mask (release film, cut with a laser cutter), which has a hole for the electrode. The flexible electrode is applied by spin coating. Finally, the mask is removed to obtain the DE with a flexible electrode shape. Two DEs are adhered to both sides of the silicone skeleton, and the frame is removed after standing for 10 hours to obtain the DEA.

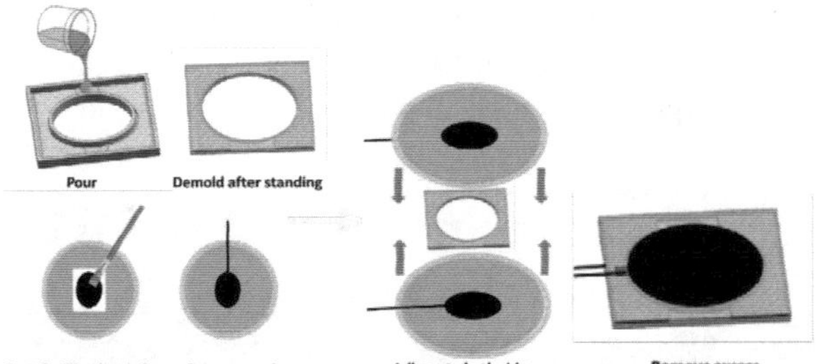

Figure 5. Fabrication process of the DEA

3.2 DEA passive tail fin

Relevant studies have shown that the swing of the passive tail fin can provide more than 90% of the propulsion force required for fish swimming, and it has the advantages of a fast swimming speed and good acceleration performance [16]. By imitating the passive caudal fin characteristics of traditional BCF fishes, a silicone passive caudal fin (Dragon Skin 30) with a similar shape is designed. The passive tail fin and the DEA are assembled together to form the DEA passive tail fin, as shown in Figure 6.

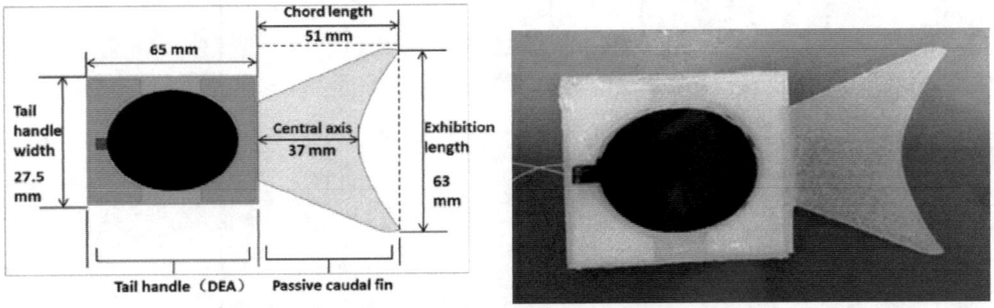

Figure 6. DEA passive tail fin

4. Results and Discussions

4.1 Displacement experiment of the DEA

A laser displacement sensor (HG-C1050, Panasonic) was used to measure the end bending displacement of the DEA. The range of actuation voltage was 3 kV-7 kV, and the frequency was 1 Hz-5 Hz. The DEs on both sides of the DEA were actuated, and both sides of the DEA could undergo bending deformation.

The end displacements of both sides of the DEA were obtained, as shown in Figure 7. The end displacements at the same frequency of the DEA are basically the same. This is because the response speed of the DEs is at the millisecond level, so there will be no obvious hysteresis at frequencies of 1-5 Hz, and the control accuracy of the DEA can be guaranteed at this frequency. With increasing actuation voltage, the end displacement of the DEA also significantly increases, and the maximum end displacement on one side is 6.48 mm. The end displacements on both sides of the DEA are basically the same under the same voltage and frequency, and the error range is within ±0.30 mm, which is mainly caused by errors in the preparation of the DEA.

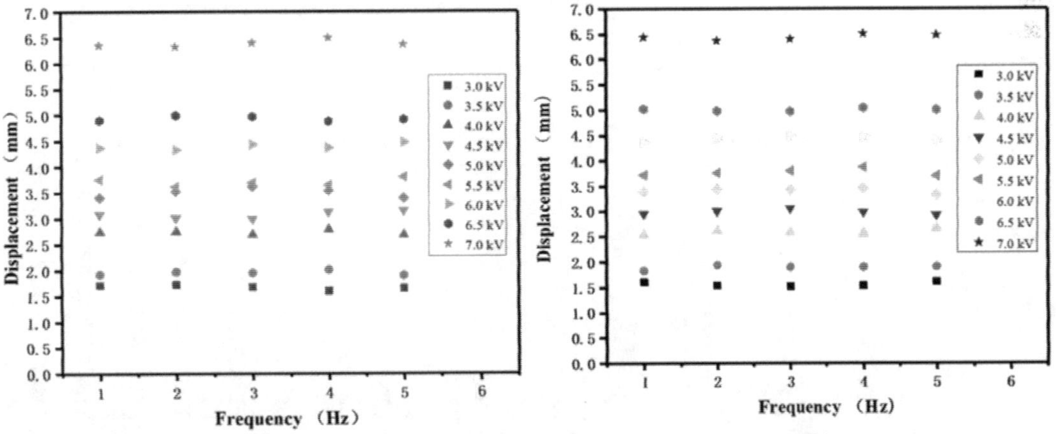

(a) (b)

Figure 7. (a) End displacement on the left side of the DEA (b) End displacement on the left right of the DEA

The end displacement data for different frequencies under the same voltage were averaged and converted into angles to obtain the relationship between the bending angle on both sides of the DEA and the voltage, as shown in Figure 8. The absolute values of the bending angle on both sides of the DEA can be added to obtain the maximum swing angle during the swing process, which can reach 31°.

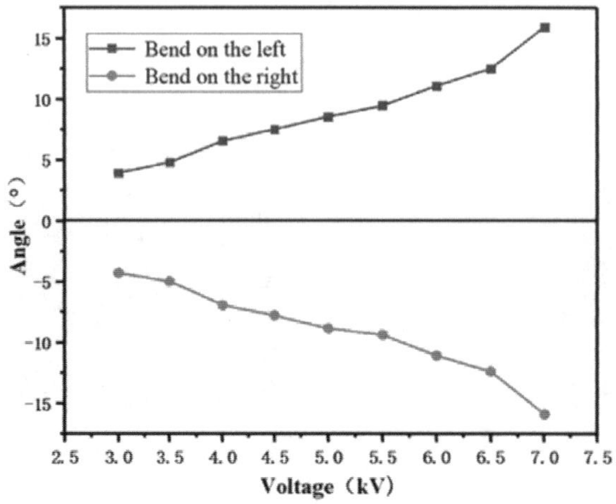

Figure 8. Relationship between the bending angle on both sides of the DEA and the voltage

4.2 Output force of the DEA

When the DEA is actuated, it provides an output force in the bending direction. In the experiment, a bridge-type force sensor was used to measure the output force provided by the DEA.

First, the effect of the actuation frequency on the output force was tested. An actuation voltage of 4 kV was input to the DE on one side of the DEA, the actuation frequency was continuously increased from 1 Hz to 5 Hz, and the output force at each time interval was measured, as shown in Figure 9. The change in the actuation frequency of the DEA has almost no effect on the output force when the actuation voltage is constant. According to the actuation characteristics of the DE, the parameters that affect the output force of the DEA include the stretch rate, actuation voltage and thickness of the DE, so the output force is constant when these parameters are constant.

Second, the effect of the actuation voltage on the output force was tested. An actuation voltage with a frequency of 1 Hz was input to the DE on one side of the DEA, the actuation voltage was continuously increased from 3 kV to 7 kV, and the output force at each time interval was measured, as shown in Figure 10. When the actuation frequency is constant, the output force of the DEA increases from 7.13 mN to 48.54 mN with increasing actuation voltage.

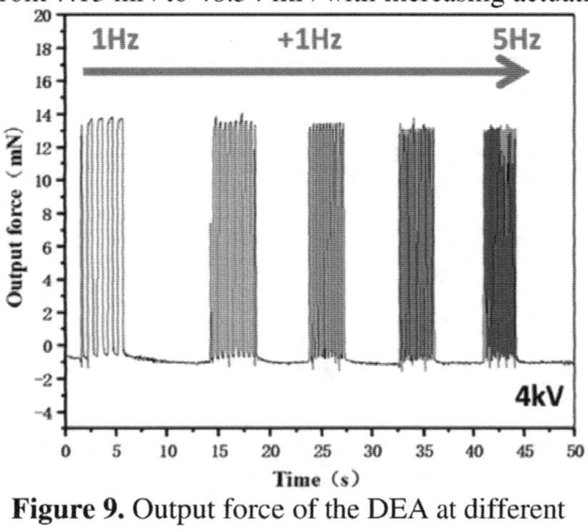

Figure 9. Output force of the DEA at different frequencies when the voltage is 4 kV

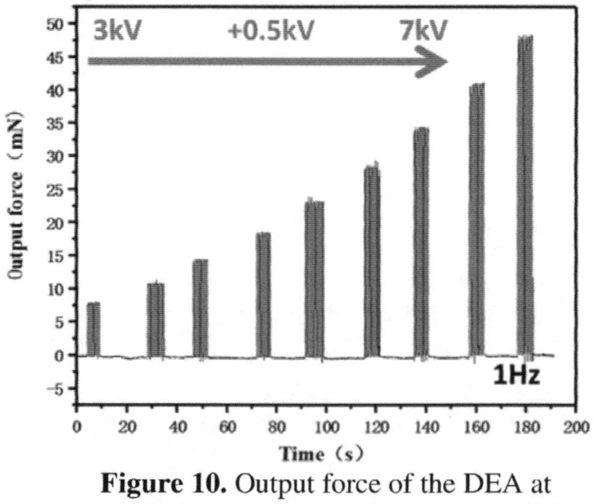

Figure 10. Output force of the DEA at different voltages when the frequency is 1 Hz

4.3 Underwater swing performance of the DEA passive tail fin

The DEA is prone to short-circuit faults when energized underwater, which can cause electrical breakdown of the DEs. To ensure safety, a safe and stable underwater energization scheme and actuation circuits were designed.

The underwater energization scheme is shown in Figure 11. Since water is a good conductor, the negative electrode (ground terminal) of the high-voltage generator is connected to the water so that the parts on the outside of the DEA that are in contact with water are all negative electrodes. The positive pole of the high-voltage generator is connected in parallel in two ways to connect the inner electrode parts of the DEA that are not in contact with water to ensure insulation of the inner and outer electrodes of the DEs. Alternate activation of the DEs on both sides of the DEA can be realized by alternately opening and closing the two positive circuits in parallel.

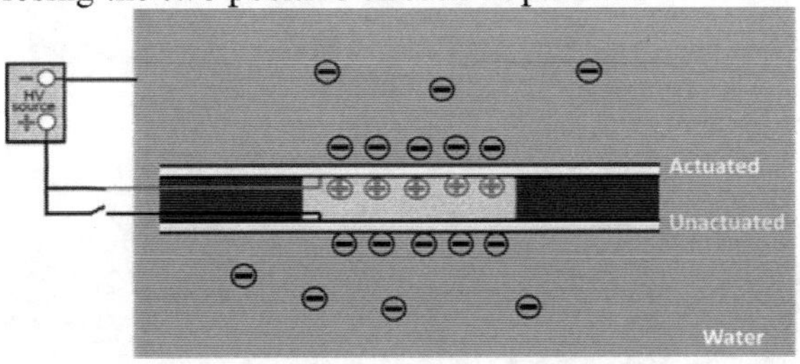

Figure 11. Underwater energization scheme of the DEA

The actuation circuit scheme uses a high-voltage generator to realize multichannel switch control at the output of the module, as shown in Figure 12. When the DEA is working, the on-off sequence of its high-voltage relay is 1 on-1 off-2 on-2 off-3 on-3 off-4 on-4 off, and the above is a cycle of swinging of the DEA. By controlling the on-off time, the law of swinging at different frequencies can be output.

When the DEA is added to the passive tail fin, in addition to the deformation caused by its own swing, it is also affected by the resistance of water. An underwater test platform was built to explore the underwater swing performance of the DEA after adding a passive tail fin, as shown in Figure 13. The DEA passive caudal fin was fixed in a water tank of 80cm×50cm×50cm, and a laser displacement sensor was used to conduct experiments on the end displacement of the DEA passive caudal fin.

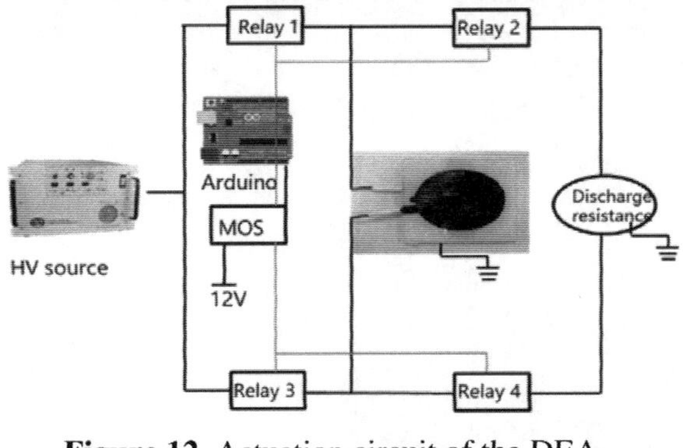

Figure 12. Actuation circuit of the DEA

Figure 13. Underwater test platform of the DEA passive caudal fin

The DEA was connected to a voltage of 3-7 kV with a frequency of 1-5 Hz. The relationship between the voltage, frequency and swing amplitude was obtained after processing the data collected by the sensor, as shown in Figure 14. The swing amplitude of the passive tail fin increases with increasing actuation voltage at the same frequency, and the swing amplitude exhibits the maximum value at 2 Hz under the same voltage. This is because the DEA has already started the next stage of actuation before the passive tail fin has reached the maximum displacement when the frequency is too high, and there is a certain hysteresis between the passive tail fin and the DEA. It reaches a maximum swing amplitude of 37.7 mm at 7 kV and 2 Hz.

The relationship between the voltage, frequency and swing speed can be obtained by processing and calculating the swing amplitude and frequency of the passive tail fin, as shown in Figure 15. The swing speed of the passive tail fin is also positively correlated with the actuation voltage at the same frequency, while it shows a maximum value at 3 Hz under the same actuation voltage. Through the analysis of Figure 14, the swing amplitude is not very different in the range of 1-3 Hz when the actuation voltage is constant. However, the difference in the cycle time is doubled at different frequencies, so the swing speed is greatly affected by the cycle time at frequencies of 1-3 Hz. It reaches a maximum value of 191.7 mm/s at 7 kV and 3 Hz.

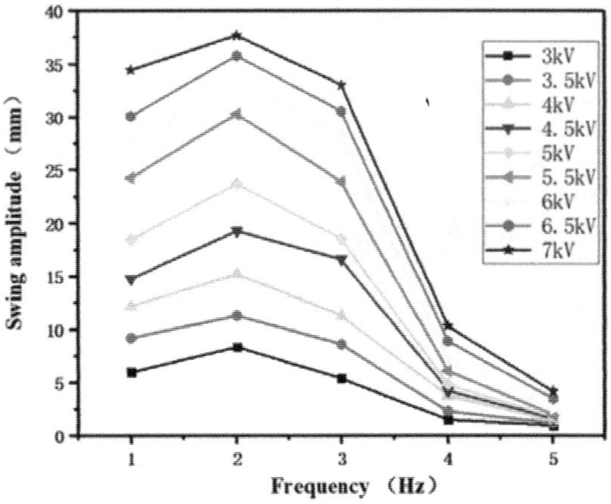

Figure 14. Relationship between voltage, frequency and swing amplitude of the DEA passive caudal fin

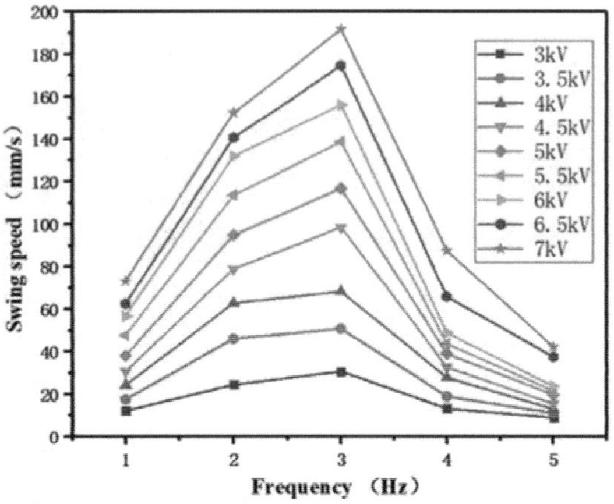

Figure 15. Relationship between the voltage, frequency and swing speed of the DEA passive caudal fin

4.4 Propulsion performance of the bionic robot fish

To explore the propulsion performance of the designed BCF-type bionic robot fish based on the dielectric elastomer actuator, the prototype was placed in still water for swimming experiments. The prototype adopted a wired connection, and the wires were drawn from the DEA and connected to a control circuit. To maintain the balance of the prototype, a balance board was added to the prototype. Voltages of different amplitudes and frequencies were applied to the bionic robot fish, and a camera was used to record its swimming process and calculate its swimming speed. Figure 16 shows the swimming process of the bionic robot fish in water at 7 kV and 3 Hz recorded by the camera.

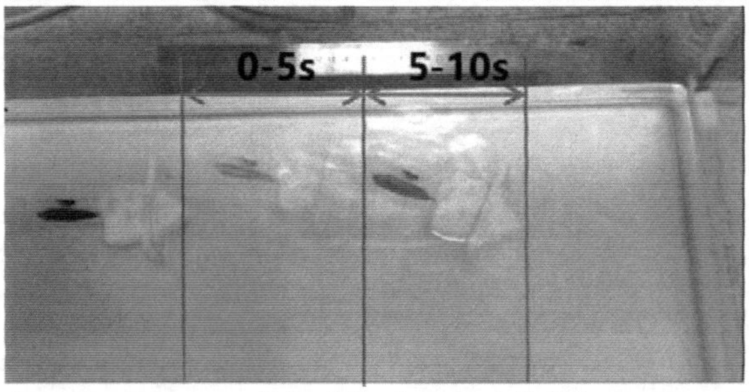

Figure 16. Swimming process of the bionic robot fish in water at 7 kV and 3 Hz

The swimming speed of the bionic robot fish at different frequencies and an actuation voltage of 7 kV was tested. The frequency was increased from 1 Hz to 5 Hz, each frequency was tested three times, and the average value was calculated. Compared with the periodic average propulsion force of the DEA passive caudal fin under each parameter in Section 4.4, the swimming speed of the bionic robot fish at different frequencies, shown in Figure 17, has a strong correlation with the periodic average propulsion obtained by CFD simulation, which also verifies the accuracy of the CFD simulation. The maximum swimming speed of the bionic robot fish is 22.7 mm/s, which is approximately 1/7 of its body length.

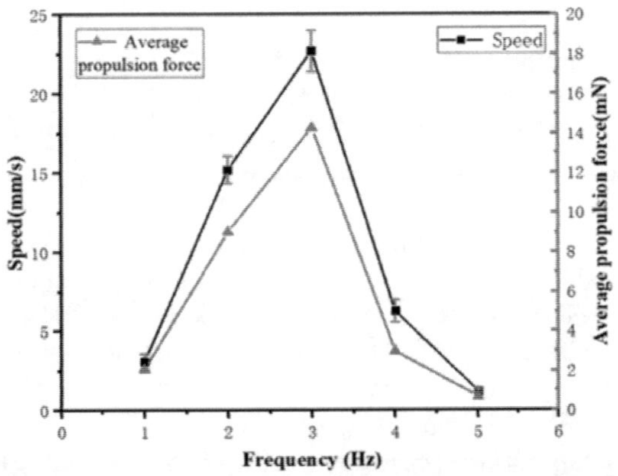

Figure 17. Comparison of speed and average propulsion force

5. Conclusions

(1) Based on the smart material of a dielectric elastomer, this study designs an antagonistic dielectric elastomer actuator according to the deformation characteristics of the DE under actuation voltage and imitation of the characteristics of the muscles of swimming fish. The DEA has the characteristics of material-structure- function integration, can realize antagonistic swing on both sides, and has the characteristics of a fast response speed, high flexibility, and easy access to the actuation source.

(2) A DEA test platform was built to conduct experiments on the end displacement and output force of the DEA, and the end displacement-voltage-frequency and output force-voltage-frequency relationships of the DEA in air were obtained. The maximum output force of the DEA is 48.54 mN. The underwater energization scheme and actuation circuit of the DEA were explained in detail, and the DEA passive tail fin was introduced to improve the propulsion performance of the DEA.

(3) The DEA was mounted on a BCF bionic robot fish, and underwater experiments were carried out. Its maximum swimming speed reaches 22.7 mm/s, which is approximately 1/7 of its body length.

6. Declaration

Funding

Supported by Beijing Nova Program of China (Grant No. 2020B00003127), Zhuhai Industrial Core and Key Technology Research Project of China (Grant No. ZH01084702180085HJL).

Availability of data and materials

The datasets supporting the conclusions of this article are included within the article.

Authors' contributions

The authors' contributions are as follows: YY and ZJ were in charge of the whole trial; HW was responsible for the design, fabrication and experimentation; HW wrote the manuscript; BL assisted with sampling and laboratory analyses. All authors read and approved the final manuscript.

Competing interests

The authors declare no competing financial interests.

References

[1] D. Rus, M.T. Tolley, Design, fabrication and control of soft robots, *Nature* **521** (2015) 467–475.

[2] M.T. Tolley, R.F. Shepherd, B. Mosadegh, K.C. Galloway, M. Wehner, M. Karpelson, R.J. Wood and G.M. Whitesides, A Resilient, Untethered Soft Robot, *Soft Robotics* **1** (2014) 213-223.

[3] M.S. Triantafyllou, G.S. Triantafyllou, An Efficient Swimming Machine, *Sci. Am.* **272** (1995) 40-48. 10.1038/scientificamerican0395-64

[4] N. Kato, M. Furushima, Pectoral fin model for maneuver of underwater vehicles, Proceedings of Symposium on Autonomous Underwater Vehicle Technology (1996) 49-56.

[5] M.S. Triantafyllou, G.S. Triantafyllou, D.K.P. Yue, Hydrodynamics of fishlike swimming, *Annu. Rev. Fluid Mech.* **32** (2000) 33-53.

[6] D. Lachat, A. Crespi, A.J. Ijspeert. BoxyBot: a swimming and crawling fish robot controlled by a central pattern generator, The First IEEE/RAS-EMBS International Conference on Biomedical Robotics and Biomechatronics (2006) 643-648.

[7] A.D. Marchese, C.D. Onal, D. Rus, Autonomous Soft Robotic Fish Capable of Escape Maneuvers Using Fluidic Elastomer Actuators, *Soft robot.* **1** (2014) 75-87.

[8] W. Coral, C. Rossi, O.M. Curet and D. Castro, Design and assessment of a flexible fish robot actuated by shape memory alloys, *Bioinspiration and Biomimetics* **13** (2018) 056009.

[9] C. Rossi, J. Colorado, W. Coral and A. Barrientos, Bending continuous structures with SMAs: A novel robotic fish design, *Bioinspiration and Biomimetics* **6** (2011) 045005.

[10] B. Kim, D.H. Kim, J. Jung, J.O. Park, A biomimetic undulatory tadpole robot using ionic polymer-metal composite actuators, *Smart Mater. Struct.* **14** (2005) 1579-1585.

[11] S. Shatara, X. Tan, E. Mbemmo, N, Gingery and S. Henneberger, Experimental investigation on underwater acoustic ranging for small robotic fish, Proceedings-IEEE International Conference on Robotics and Automation (2008).

[12] Z. Chen, T.I. Um, H. Bart-Smith, Ionic polymer-metal composite enabled robotic manta ray, Electroactive Polymer Actuators and Devices, 2011.

[13] Z. Zhang, M. Philen, W. Neu, A biologically inspired artificial fish using flexible matrix composite actuators: analysis and experiment, *Smart Mater. Struct.* **19** (2010) 094017.

[14] T. Li, G. Li, Y. Liang, T. Cheng, J. Dai, X. Yang, B. Liu, Z. Zeng, Z. Huang, Y. Luo, T. Xie and W. Yang. Fast-moving soft electronic fish. *Sci. Adv.* **3** (2017) e1602045. 10.1126/sciadv.1602045

[15] G. Li, X. Chen, F. Zhou, Y. Liang, Y. Xiao, X. Cao, Z. Zhang, M. Zhang, B. Wu and S Yin, Self-powered soft robot in the Mariana Trench, *Nature*, **591** (2021) 66–71.

[16] X. Ren, J. Lou, Z. Jia, Y. Yang, T. Chen, H. Meng, Oscillating Performance and Fluid-structure Interaction Mechanism of a Small Koi's Caudal Fin-like Underwater Propulsion Actuated by MFC, *Journal of Mechanical Engineering* **55** (2019) 214-221, 230.

[17] R. Pelrine, R. Kornbluh, Q. Pei, J, Joseph, High-speed electrically actuated elastomers with strain greater than 100%, *Science* **287** (2000) 836-839.

[18] R. Pelrine, R. Kornbluh, G. Kofod. High-Strain Actuator Materials Based on Dielectric Elastomers. *Adv. Mate.* **12** (2000) 1223-1225.

[19] S. Wang, B. Huang, D. Mccoul, M. Li, L. Mu, J. Zhao, A soft breaststroke-inspired swimming robot actuated by dielectric elastomers, *Smart Mater. Struct.* **28** (2019) 045006.

[20] T. Cheng, G. Li, Y. Liang, M. Zhang, B. Liu, T. Wong, J. Forman, M. Chen, G. Wang, Y. Tao and T. Li, Untethered soft robotic jellyfish, *Smart Mater. Struct.* **28** (2018).

[21] A. Raj, A. Thakur, Fish-inspired robots: design, sensing, actuation, and autonomy-a review of research, *Bioinspir. Biomim.* **11** (2016) 031001.

[22] M. Sfakiotakis, D.M. Lane, J.B.C. Davies, Review of fish swimming modes for aquatic locomotion, *IEEE J. Oceanic Eng.* **24** (1999) 237-252.

Numerical Investigation on the Atomization Characteristics of An Aerosol Spray Nozzle

Liang Liu[1,a], Xingwei Wu[2,b], Dian Wang[3,c] and Yue Yin*[4,d]

[1] Marine design & research institute of China, Shanghai, China
[2] Marine design & research institute of China, Shanghai, China
[3] College of Power and Energy, Harbin Engineering University, Harbin, China
[4] College of Power and Energy, Harbin Engineering University, Harbin, China

[a] 675912732@qq.com, [b] xingwei_wu@163.com, [c] wangdian20000508@163.com, [d] yinyue@hrbeu.edu.cn

Abstract. Inlet air filtration of ships and coastal installations is mainly used to filter out salt aerosol particles, so researches on the inlet filter needs to simulate spray aerosol state in the marine environment. The sea salt aerosol particles have multiple sizes, thus salt aerosol generator should have the ability to produce salt aerosol particles of various diameters. This paper investigates the factors that affect the particle diameter of a salt aerosol generating device. Numerical simulations on the two-phase flow in a salt aerosol nozzle are conducted and the influence of air inlet pressure and droplet mass flow rate on the particle diameter is discovered. The results show that the particle diameter of the aerosol nozzle decreases with the air inlet pressure increasing and the droplet mass flow rate decreasing.

1. Introduction

There is salt in the ocean atmosphere and it exists in the form of aerosol. There are many large surface ships around the world equipping with gas turbine power, and the large intake volume of gas turbine results in a higher demand on the filter of salt aerosol, which requires the designer to test the desalination performance of marine filtration devices in the marine environment. It is the aim of this paper to provide better ocean state simulation for the designed products. In the development of marine filter, it is necessary to simulate the real working environment by spraying aerosols of various aerosol particle diameters. Kumar [1] developed a mathematically model to dynamically predict the droplet diameter and the droplet residence time of a nozzle that atomizes the solution (pharmaceutical compound and organic solvent) in an antisolvent environment. The author studied the dependencies of droplet size on process parameters such as nozzle diameter and solution flow rate. The results show that the droplet diameter firstly develops to the maximal value, then it starts to decrease and gradually reaches zero as the solvent from droplet totally evaporates. It is also found that the maximum droplet diameter increases with the nozzle diameter increasing and decreases with the solution flow rate increasing; the droplet residence time decreases with the solution flow rate increasing or the initial droplet diameter decreasing. Han et al. [2] numerically and experimentally investigated the atomization characteristics of an internal mixing air atomizing nozzle for dust reduction. The focus is on discovering the effect of the water supply pressure on the atomization characteristics of the nozzle. The results showed that the increase of water supply pressure would increase the water flow velocity and the internal pressure in the mixing chamber of the nozzle, meanwhile, the water flow rate exponentially increases and the air flow rate exponentially decreases. It is also observed that the droplet size and velocity both increase with the water supply pressure. Lastow and Balachandran [3] experimentally investigated the spray and atomization properties of a low voltage electrohydrodynamic nozzle using the PDA instruments. The results show that the operation voltage range of water is much wider than that of the saline. The saline aerosol and the water aerosol had a mean droplet diameter of 1 mm and 2.1–2.9 mm separately. The droplet sizes remain constant with the increase of the liquid flow rate since the flow rate used in the experiments is not in the range for which the scaling laws are applicable. The authors also conducted numerical simulations on the electrohydrodynamic atomization of water and found good agreements of the numerical results with the experimental results such as the shape and angle of the sprayed cone jet and the droplet sizes. Hii et al. [4] numerically and experimentally studied the impinging aerosols method for encapsulation in which the two materials are alginate and $CaCl2$. The results show that a higher flow rate results in smaller droplet diameters but a wider range of droplet size. The increase of the fluid viscosity resulting from the increase of alginate concentration leads to larger droplets and smaller spray angles. Verma et al. [5] numerically modelled the dynamics of the aerosols from an internal mixing nozzle using different multiphase models including DPM (discrete phase model) and VOF (volume of fluid). The authors concluded that the mean diameter of the droplets decreases but their velocity and number increase with the air inlet pressure increasing. And the air inlet pressure presents more influence on the features of the droplets than the water flow rate. Charras et al. [6] developed a mathematical model to predict the particle diameter distribution of the aerosols produced by the pressurized metered dose inhalers. It was found that the particle diameter distribution can be mathematically represented by a lognormal function with constant geometric standard deviation. Jiang et al. [7] reviewed the advanced simulation methods of gas-liquid two-phase flows for atomization and sprays typically in combustion engines. The authors summarized the application situations, basic equations and computational costs of the numerical methods for gas-liquid two-phase flows including the applications of LES and DNS in simulating the atomization and spray processes. In this paper, with a given nozzle as a model and some fixed parameters, the external flow field is studied by numerical simulations, and the factors affecting the particle diameters of the salt aerosol generating device are investigated.

2. Computational Setup

2.1. Geometries

The three-dimensional model of the aerosol spray nozzle is established on the software GAMBIT. Figure 1 shows the schematic of the aerosol spray nozzle and its computational domain. The diameter of the central droplet inlet is 1.3mm. The surrounding cylindrical atmosphere has a height of 1000mm and a diameter of 100mm.The four vents next to the central droplet inlet and the annular hole are the inlets of the high-pressure gas, and the two bent converging pipes are also for ejecting the high-pressure gas. The droplets come out from the central hole due to gravity, and are broken up into smaller droplets by the high-pressure high-velocity gas flowing from the four vents and the annular hole. The tiny droplets are further broken up into aerosols by the high-pressure high-velocity gas flowing from the two bent converging pipes. Hence the high-pressure gas plays the role of atomization and ejection.

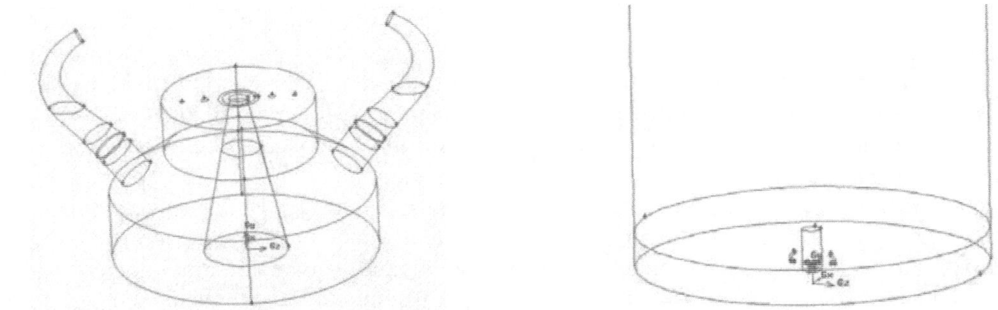

(a) Schematic of the interior and exterior flow paths (b) Schematic of the overall model

Figure 1. The designed geometrical structure and computational model of the aerosol spray nozzle

3. Numerical Methods and Boundary Conditions

The computational domains were discretized by tetrahedral grids and the total element number is about 90 thousand. Figure 2 shows the computational grids for the aerosol spray nozzle in overall and in detail. The mesh near the droplet inlet and the high-pressure gas inlets are refined. The numerical simulations are conducted on the platform FLUENT19.0. The air inlets are specified as the pressure inlet. The droplet inlet is specified as mass flow inlet. And the cylindrical surface of the surrounding atmosphere is specified as pressure outlet. The flow in the aerosol spray nozzle is typically two-phase flow and is handled with as a grain flow. The strategy is to solve the gas phase background flow with RANS methods in conjunction with the standard k-ε turbulence model. The gas phase is assumed as ideal gas and the two-phase flow is assumed as compressible flow. The convective terms are discretized with the second upwind scheme and the dissipation terms are discretized with the central difference scheme. The pressure-velocity coupling scheme is SIMPLE algorithm. Moreover, the motion of droplets and particles is simulated by the discrete phase model (DPM). The random orbit model is adopted as the turbulence interference model for particle flow. The DPM model is calculated every ten iterations of the gas phase calculation. The diameters of the scattered particles are worked out under different gas inlet pressures and droplet mass flow rates.

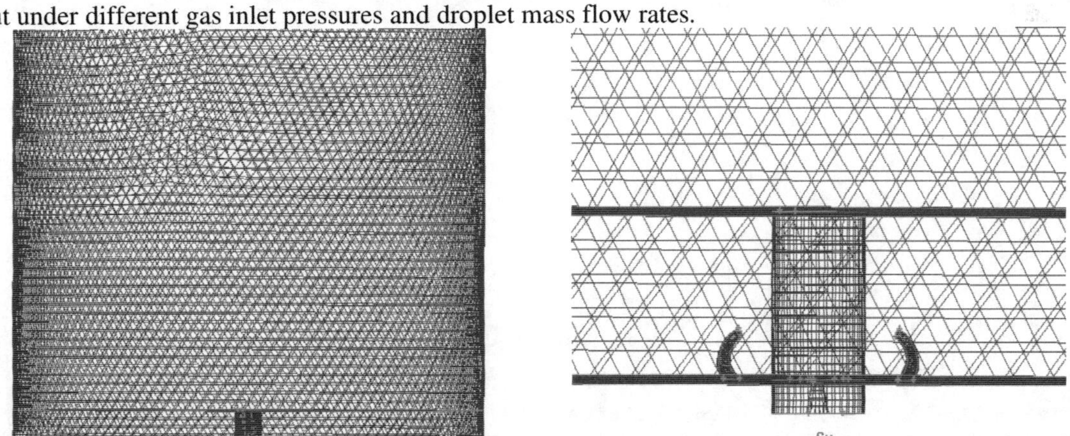

(a) Overall view (b) Detailed view

Figure 2. Computational grids for the aerosol spray nozzle

4. Results and discussion

4.1. Effect of droplet mass flow rate

In this section, four cases with different droplet mass flow rate and the same air inlet pressure are simulated and compared. The initial particle diameter and velocity are 0.03mm and 10m/s. The values of droplet mass flow rate are 0.6g/s, 0.8g/s, 1g/s and 1.5g/s. The air inlet pressure keeps constant at 105325Pa. Figure 3-6 presents the results of velocity vectors, velocity contours and particle diameter distribution of the four cases on the central slices. When the droplet mass flow rate is 0.6g/s, the maximum velocity magnitude of the ejected gas is 77.1m/s and the particle diameter of the sprayed aerosol is in the range of 0.03-0.06mm. When the droplet mass flow rate is 0.8g/s, the maximum velocity magnitude of the ejected gas is still 77.1m/s and the particle diameter of the sprayed aerosol is in the range of 0.03-0.0723mm. When the droplet mass flow rate is 1g/s, the maximum velocity magnitude of the ejected gas is unchanged and the particle diameter of the sprayed aerosol is in the range of 0.03-0.074mm. When the droplet mass flow rate is 1.5g/s, the maximum velocity magnitude of the ejected gas is unchanged and the particle diameter of the sprayed aerosol is in the range of 0.03-0.0828mm. Therefore, the particle diameter of the investigated aerosol spray nozzle increases with the droplet mass flow rate.

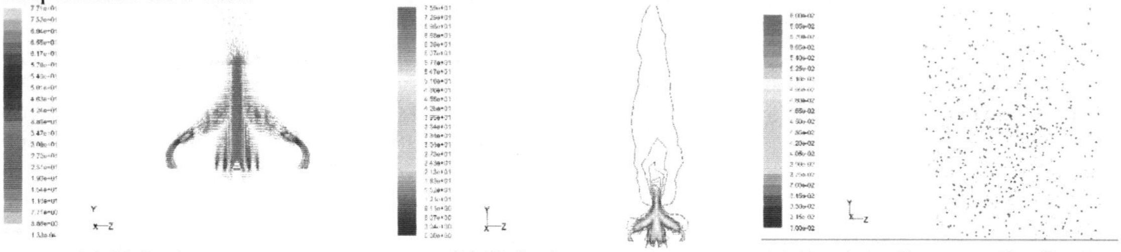

(a) Velocity vectors (b) Velocity contour (c) Particle diameter distribution

Figure 3. The numerical results of aerosol spray nozzle, $\dot{m}$ =0.6g/s

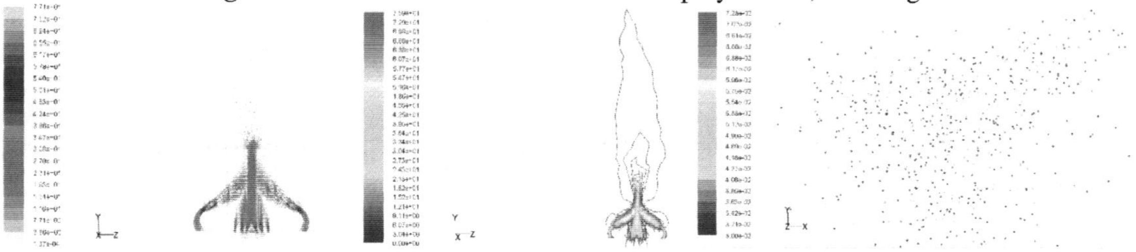

(a) Velocity vectors (b) Velocity contour (c) Particle diameter distribution

Figure 4. The numerical results of aerosol spray nozzle, $\dot{m}$ =0.8g/s

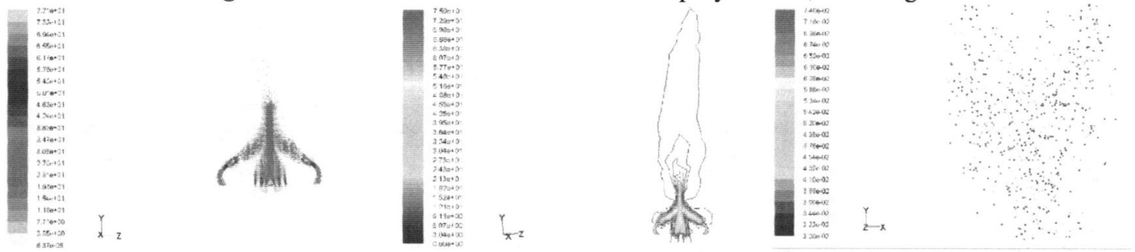

(a) Velocity vectors (b) Velocity contour (c) Particle diameter distribution

Figure 5. The numerical results of aerosol spray nozzle, $\dot{m}$ =1g/s

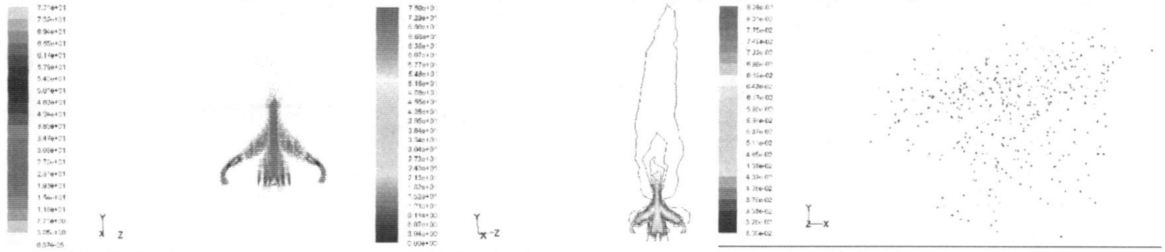

(a) Velocity vectors (b) Velocity contour (c) Particle diameter distribution

Figure 6. The numerical results of aerosol spray nozzle, $\dot{m}$ =1.5g/s

5. Effect of air inlet pressure

In this section, three cases with different air inlet pressure and the droplet mass flow rate are simulated and compared. The initial particle diameter and velocity are 0.03mm and 10m/s. The values of air inlet pressure are 104325Pa, 105325Pa and 109325Pa. The droplet mass flow rate keeps constant at 0.8g/s. Figure 7-9 presents the results of velocity vectors, velocity contours and particle diameter distribution of the three cases on the central slices. When the air inlet pressure is 104325Pa, the maximum velocity magnitude of the ejected gas is 66.8m/s and the particle diameter of the sprayed aerosol is in the range of 0.03-0.102mm. When the air inlet pressure is 105325Pa, the maximum velocity magnitude of the ejected gas is 77.1m/s and the particle diameter of the sprayed aerosol is in the range of 0.03-0.0723mm. When the air inlet pressure is 109325Pa, the maximum velocity magnitude of the ejected gas is 108 m/s and the particle diameter of the sprayed aerosol is in the range of 0.03-0.0624mm. Therefore, with the air inlet pressure increasing, the velocity of the ejected gas increases, the sprayed particle diameter decreases and the atomization performance of the aerosol spray nozzle is improved.

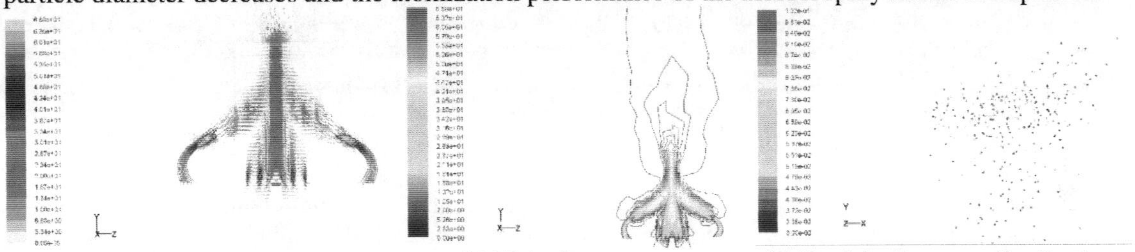

(a) Velocity vectors (b) Velocity contour (c) Particle diameter distribution

Figure 7. the numerical results of aerosol spray nozzle, P=104325Pa

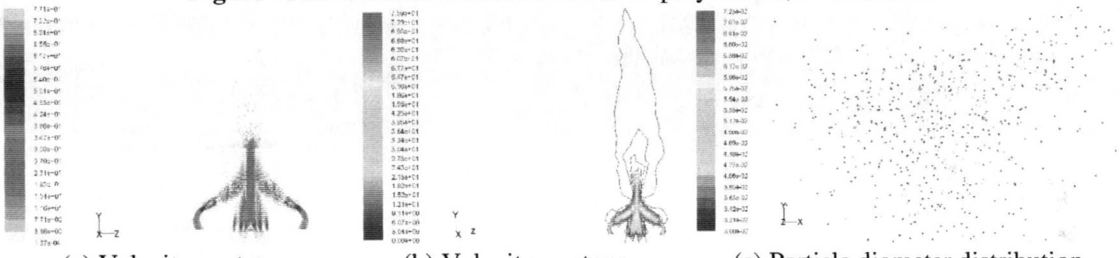

(a) Velocity vectors (b) Velocity contour (c) Particle diameter distribution

Figure 8. the numerical results of aerosol spray nozzle, P=105325Pa

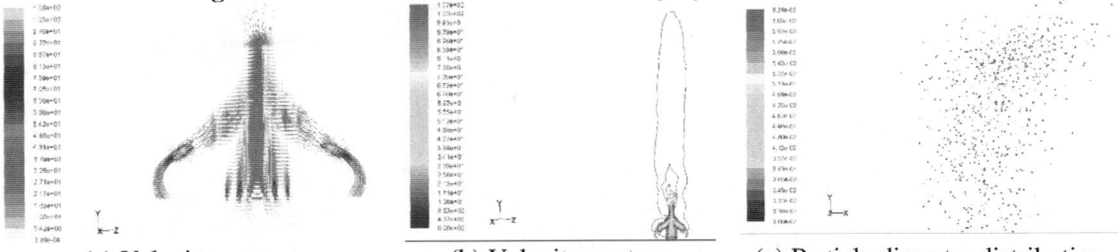

(a) Velocity vectors (b) Velocity contour (c) Particle diameter distribution

Figure 9. the numerical results of aerosol spray nozzle, P=109325Pa

6. Conclusions

In this research, the effects of air inlet pressure and droplet mass flow rate on the atomization performance of a designed aerosol spray nozzle are numerically investigated. The numerical model is established and spatially discretized in GAMBIT. The numerical simulations are carried out on FLUENT. The standard k-ε model is selected as the turbulence model. The motion of droplets and particles is simulated by the discrete phase model (DPM). The velocity distribution of the gas phase and the particle diameter distribution of the water phase are displayed and analyzed. The results show that the particle diameter of the investigated aerosol spray nozzle increases with the droplet mass flow rate. And the velocity of the ejected gas increases, the sprayed particle diameter decreases and the atomization performance of the aerosol spray nozzle is improved with the air inlet pressure increasing.

7. References

[1] Kumar R 2022 Numerical simulation to estimate the droplet size in aerosol solvent extraction system *Materials Today: Proceedings* **57** pp 1515-1519

[2] Han H, Wang P, Li Y, Liu R and Tian C 2020 Effect of water supply pressure on atomization characteristics and dust-reduction efficiency of internal mixing air atomizing nozzle *Advanced Powder Technology* vol 31(1) (Netherlands: Elsevier B V) pp 252-268

[3] Lastow O and Balachandran W 2007 Novel low voltage EHD spray nozzle for atomization of water in the cone jet mode *Journal of Electrostatics* vol 65(8) (Netherlands: Elsevier B V) pp 490-499

[4] Hii Y S, Law M C and Chan Y S 2021 Experimental and numerical study of the impinging aerosols method for the micro-encapsulation of phosphate solubilising microorganisms (PSMs) *Biochemical Engineering Journal* vol 174 108118 (Netherlands: Elsevier B V)

[5] Verma N, ManojKumar K and Ghosh A 2017 Characteristics of aerosol produced by an internal-mix nozzle and its influence on force, residual stress and surface finish in SQCL grinding *Journal of Materials Processing Technology* vol 240 (Netherlands: Elsevier B V) pp 223-232

[6] de Charras Y L, Ramírez-Rigo M V and Bertin D E 2022 Prediction of the particle size distribution of the aerosol generated by a pressurized metered-dose inhaler *Powder Technology* vol 399 117151 (Netherlands: Elsevier B V)

[7] Jiang X, Siamas G A, Jagus K and Karayiannis T G 2010 Physical modelling and advanced simulations of gas–liquid two-phase jet flows in atomization and sprays *Progress in Energy and Combustion Science* vol 36(2) (England: Elsevier B V) pp 131-167

Design of a Novel Wireless NoC Architecture for Chip Multiprocessor

Yin Xin, Guozhi Song*

TianGong University, Tianjin, China

Corresponding author's e-mail: songguozhi@tiangong.edu.cn

Abstract. The Network on chip is a new communication method of System on Chip. It is the main part of multi-core technology. A large number of embedded cores can be combined on a single chip. Due to the multi-hop characteristics of communication, there are some problems in the design of traditional NoC, such as high latency and power consumption. Thus, to handle the communication problem of long-distance transmission of traditional NoC, wireless Networks on Chip (WiNoC) has been proposed. Using WiNoC to design the multi-core System on Chip can significantly reduce the latency and energy consumption of the network. The advantage of using wireless network on chip is to replace the multi-hop path of traditional network on chip with wireless single hop link. However, due to the limited space on the chip and the cost, it is very important to determine the optimal number and correct location of wireless hubs. In this paper, we propose a new method to obtain the optimum configuration of a WiNoC by using the Co-evolutionary Algorithms (CA) and a wireless NoC architecture based on mesh. This work examines the effect of VC and sub-net size on WiNoC performance efficiency and energy under different packet injection rate (PIR) and the number of cores (64, 144 and 256). The simulation experiment of the algorithm proposed in the whole architecture has been implemented on the Noxim platform.

1. Introduction

With the development of semiconductor technology, the size of electronic components is getting smaller and smaller, which requires the continuous improvement of chip integration. More and more IP cores need to be integrated on a single chip[1]. At present, a large number of IP cores on a single die that is between hundred and thousand. According to the forecasting of the International Road-map of Semiconductor, the number of integrated processing elements in a chip will reach more than 2000 before the end of the next decade[2]. The traditional bus structure has become a bottleneck restricting the development of chip size, speed, power consumption, and communication requirements. The recently emerging communication backbone is the NoC. The emergence of NoC has good scalability and reusability. While supporting parallel communication, it realizes the clock control of global asynchronous and local synchronization, and solves the single clock problem caused by sharing buses. The main idea of NoC is to apply the network technology to the chip design, so that the data transmission between IP cores can be carried out through the router[3]. Despite the advantages of NoC, when the network on chip structure becomes larger and larger, the IP core (Intellectual Property core) at the edge and diagonal position is farther away, and the number of forwarding hops increases during communication, which will cause transmission delays. Time and power consumption increase. The traditional wired NoC is unable to meet the needs of communication and the need for increased IP cores for chip development due to the limitation of planar metal interconnection wiring. When the network size increases. The conventional NoC has high communication latency and power consumption due to their multi-hop characteristics of long-distance communication. To alleviate this problem, using wireless RF(radio frequency) [4] have been proposed. It can improve the latency and power consumption. In a Wireless Network-on-Chip (WiNOC), the wired multi-hop path between long-distance IP cores is replaced by single hop wireless link. In this way, the reduction of hops will reduce latency and power consumption. Therefore, WiNoC has better performance and this architecture is a good alternative to face the problem of data transmission [5,6].

In this thesis, we propose a hybrid wireless Network on Chip architecture. The architecture of WiNoC is divided into several smaller, interconnected core clusters called sub-nets. Each sub-net is composed of several IP cores. Some IP cores are connected to a radio hub with a wireless interface, and they can be wireless connected to the radio-hubs (RH) of all sub-nets in the network. When the nearby cores communicates, it uses the traditional wired. But when the long-distance nodes communicates, it is mainly realized through the wireless link. By using this hybrid method and selecting the best number and location of wireless hubs, the performance of the network can be significantly improved. First, we determine which hubs should have wireless interfaces. The optimal placement of the wireless hub is done through the co-evolution algorithm (CA) algorithm.

This paper includes five parts. The next section introduces the related work of this paper. The third section introduces the architecture design of this paper, mainly including: the design details of the hybrid architecture based on wired and wireless interconnection, the description of the number and placement of WIs, and the experimental design process of the simulation platform. The four section introduces the simulation results and discussion. Finally, some conclusions and future work are given.

2. Related works

WiNoC has become a research hotspot in recent years. It is considered to solve the scalability limitations of traditional multi-hop NOC architectures. Below, some WiNoC architecture are briefly presented. In WiNoC, there has been a lot of work on different topology. D Zhao et al. [7] proposed a pure wireless Network on Chip architecture. All nodes in the architecture are wireless nodes. Any two adjacent nodes can communicate through wireless links to form a two-dimensional pure wireless Mesh structure, but this structure uses wireless multi-hop communication, resulting in relatively large power consumption and area of the entire network. A Ganguly et al. [8] designed a multi-core platform wireless on-chip network architecture, using the advantages of wired and wireless communications, developed a hybrid mechanism for data transmission between IP cores, and provided on-chip wireless communication through on-chip antennas. Transmit data over long distances. C Wang et al. [9] proposed a hybrid WiNoC architecture called the Wireless NePA structure, in which all nodes are connected by wired links to form a two-dimensional wired network on chip, and then the entire network is divided into four 5×5 sub-modules of the same size. Network, the center of each sub-net is set as a wireless node. Set up wireless links between wireless routers to enable wireless communication between wireless routers in each sub-net. The problem with this structure is that with the gradual increase of the network scale, the routing distance of the communication within the sub-net also increases, so a certain time delay will occur.

In this regard, we propose a architecture NoC based on mesh topology. This work introduces a comprehensive design methodology of a hierarchical WiNoC.

3. Methodology

3.1. Design of WiNoC architecture

We consider the WiNoC architecture. If the number of WI is too large, the cost of wireless will be higher. If the number of WI is too small, the communication latency and energy consumption can not be well alleviated. Therefore, we should strategically place a limited number of wireless interfaces (WI) to obtain the best performance. First, we divide the whole topology into several adjacent small clusters, that is, sub-nets. It consists of multiple cores, allowing cores belonging to a specific sub-net to communicate with distant sub-nets [10]. In the WiNoC architecture, each sub-net is a connection, in which several sets of cores are connected to a radio hub. As described in section 3.2, we use simulated annealing to optimize the placement of WI.

Similar to the other NoC, the data packets are divided into smaller parts called flits [11].

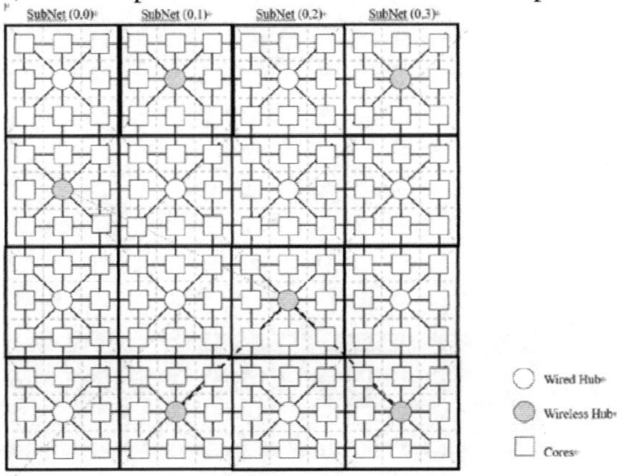

Figure 1. The optimum position of the wireless nodes for the network with 16 hubs at the second level

The proposed Hybrid NoC architecture is with both wired and wireless interconnects. As it is shown in the Figure 1, there are three different classifications of interconnections. First, Core-Core and Core-Hub interconnections: a wired point-to-point connection between two IP cores as well as between IP core and a hub. The third interconnect category is Hub-Hub connection between radio-hub elements. These connections can be either wired or wireless.

As can be seen in Figure 1, the topology is based on 12×12 as an example, divided into 16 sub-nets, and every 9 adjacent cores form a sub-net. The nodes in each sub-net are connected to a hub [12]. Therefore, there are 16 hubs connected together. The main advantages of this topology are:

(1) In this structure, the communication between the certain sub-nets is wireless, which can reduce hops in the transmission between the sub-nets and ease the communication pressure.
(2) This topology has fewer wireless hub devices and can control the area overhead of the network. When there is a lot of data transmission in the network, the data can be shunted, effectively reducing the network latency.
(3) The topology has good scalability. With the increase of the number of IP cores on the chip and the topology size changes, just add nodes according to the specific description rules in this paper.

3.2. Optimizing the number of WIs and placement

To determine the optimal quantity and location of WIs in a network and the best position of WIs has a significant impact on the network performance. We used co-evolutionary algorithms (CA) [13]. We choose the CA to solve this compromise optimization problem because it is robust for solving complex nonlinear system optimization problems. As an optimization algorithm, it simulates an ecosystem composed of two or more species. Multiple species in the ecosystem co-evolve and interact, leading to the sustainable evolution of the ecosystem [14]. CA is famous for its fast convergence[15]. Moreover, CA considers the co-evolution of sub-populations of individuals representing specific parts of the global solution.

Firstly, we assign the Qi probability of having a wireless interface to the i-th node. Let N be the number of hubs of the network. Let d be an N*N matrix, where element dij is the distance (shortest path) between hub i and hub j measured in hops. The matrix d is populated using Dijkstra's shortest path algorithm [16]. Qi can be calculated as follows:

$$Qi=\frac{\sum_{j=1}^{N} dij}{\sum_{i=1}^{N}\sum_{j=1}^{N} dij} \quad (1)$$

It can be seen that Qi is directly proportional to the sum of the distance (which is calculated in the number of hops) between the i-th node and other nodes. Evidently, the higher the sum, the greater the probability.

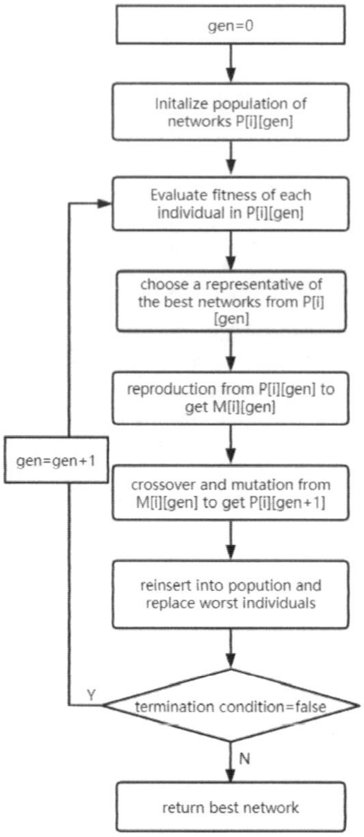

Figure 2. Flowchart illustrating the co-evolutionary algorithms we used for the network optimization

Initially, WI is randomly placed,and the probability of obtaining WI is equal for each center. CA is executed on the optimization metric pi defined by "(1)".

We use a standard CA, as shown in the flowchart of Figure 2. According to the calculated probabilities in "(1)".

The optimization process of CA is used to select the best individual from each species as the representative of the species to make it converge quickly [17]. When a new network changes its decision variables to obtain better fitness value, it will change the system response in turn, changes the fitness value of the other network. The other network will behave in the same way. The adjustment process will continue until no one can gain better fitness value.

3.3. Design flow of NoC on Noxim

This section described the experimental setup. This experiment was simulated using Noxim[18]. The operating system is Ubuntu 14.04. This paper chooses three network sizes: 8x8, 12x12 and 16x16 (64, 144 and 256 nodes respectively).

Table 1 outlines the proposed Hybrid mesh NoC configurations in this paper.

Table 1. NoC Configuration

Parameter	**Value**
Network size	12*12
Number of radio-hubs	6
Switching technique	Wormhole
Number of Channels	2,4,8
Wireless data rate (Gbps)	16
Flit size(bit)	32
Simulation(cycles)	10000
Radio Access Control	Token Packet
Selection Strategies	Random
Technology	65nm
Clock Frequency (Ghz)	1

Table 1 shows the Noxim simulation setup used in the research. It randomly distributes data packets from a source to a destination, and each node communicates effectively by sending data packets to each other with the same probability. In this study, we evaluated the performance impact in the WiNoC architecture in terms of average network latency, network throughput, energy consumption and average wireless utilization. The experimental conditions are set as follows:

(1) Packet Injection Rate (PIR): The probability that a data packet is injected into the on-chip network. For any single-node router in the on-chip network, it is the standard number of data packets sent in each clock cycle. When PIR=0.001 means that the node sends 0.001 data packets in 1 clock cycle (that is, 1 data packet is sent every 1000 cycles). The PIR studied in this paper are 0.001-0.03.

(2) The three different network sizes are divided into 4×4 sub-nets, and there are 6 wireless hubs. For an 8×8 network topology, there are 4 cores in each sub-net. For a 16×16 network topology, there are 16 cores in each sub-net.

(3) Traffic distribution strategy: random and uniform distribution, that is, each source node in the network randomly sends data to other nodes with the same probability, and the destination node for each transmission is randomly generated.[19] It can show the data transmission situation of the network in real-time communication state, and reasonable data and conclusions can be obtained.

4. Results and discussion

In this section, we analyze the performance of Noxim testing the effect of an increasing PIR (Packet Injection Rate) on throughput dynamic energy and latency. We analyze the effect of different variety of workloads using several numbers of radio hubs: the effect of an increasing PIR on throughput dynamic energy and latency.As we all know, the main factors that affect the performance of WiNoC are sub-nets, virtual channels and routing algorithms. Therefore, in addition to the above experiments, we also specifically analyzed the impact of virtual channels on throughput, energy consumption and delay in the architecture we designed.

4.1. Throughput

The network throughput is an important index to test network performance, which can be used to measure the performance efficiency and consistency of network connection. That is, the rate at which the network successfully receives data represents the ability of the network to carry and transmit data. Throughput T is defined as equation (2), where Rflits represents the number of microchips received by the network, Nnodes represents the total number of nodes, and Ncycles represents the total clock cycles from the generation of the first flit to the receipt of the last flit.

$$T=R_{flits}/(N_{nodes} \times N_{cycles}) \tag{2}$$

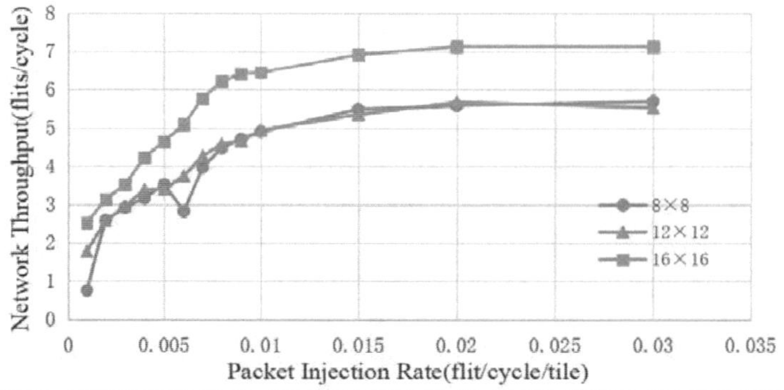

Figure 3. The network throughput vs. PIR for the number of six radio hubs as function as different network topology

Figure 3 reveals a comparison of the network throughput of three different sub-net sizes and variable PIR. The smaller the sub-net size is, the lower the network throughput changes with PIR. It can be seen from the figure that when the PIR is less than 0.015, there is no congestion in the three topologies, and the throughput of 8×8 and 12×12 is basically the same; the network size is 16×16, that is, the average network throughput when the sub-net size is 4×4 is better than the other two network sizes.

4.2. Dynamic energy

The energy consumption in the NoC refers to the energy consumption caused by processing and transmitting data in the network per unit time.

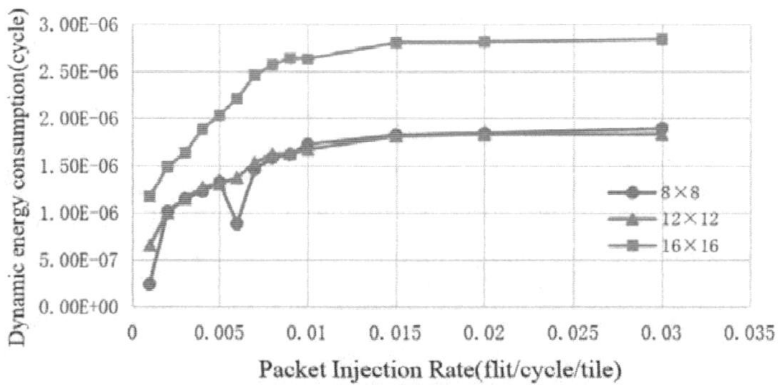

Figure 4. The dynamic energy vs. PIR for the number of six radio hubs as function as different network topology

Figure 4 show the estimated trend-lines of the dynamic energy vs. PIR for 8 x 8, 12 x 12 and 16 x 16. It is possible to observe that 16 x16 lost completely with higher energy dissipation for the three different topology. From Figure 4, the lowest energy dissipation is achieved by using 8 x 8 when the PIR vary from 0.001 to 0.015. The lowest energy consumption of 8 x 8 is due to the fact that the packets in this topology reach the destination node and the wireless nodes pass through less when communicating.

4.3. Latency

The network latency mainly refers to the time spent in the whole process of transmitting a packet. It starts from the source node, passes through multiple intermediate nodes and finally reaching the destination node. The network latency is usually in milliseconds. The network latency reflects the data transmission rate of the network. It is considered to be the number of clock cycles lost between the time when the first microchip is injected into the network from the source node and the time when the destination node receives the last microchip. It includes transmission delay, transmission delay and receive delay. Average network latency Daverage is defined as equation (3), which is the numerical average of the total delay; N is the number of data packets that reach the target node information within a certain period of time; Di is the delay generated by the i-th node. Under normal circumstances, the smaller the average delay, the better the network performance.

$$D_{average}= \frac{1}{N}\sum_{i}^{N} Di \quad (3)$$

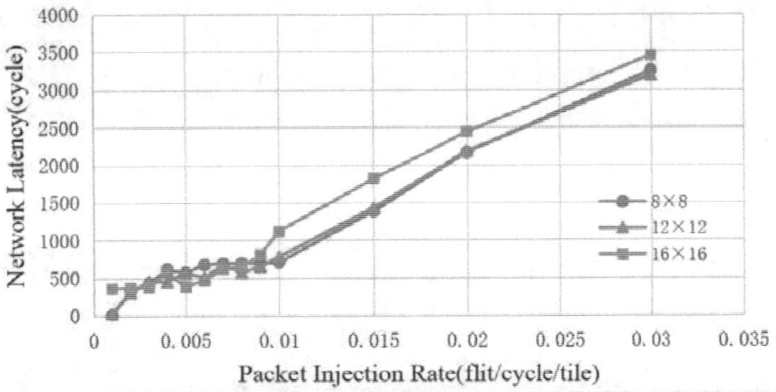

Figure 5. The network latency vs. PIR for the number of six radio hubs as function as different network topology

Figure 5 shows the network latency comparison of three different sub-net sizes and variable PIR. It can be seen from the figure that when the PIR is less than 0.009, the network latency of these three topologies are small and basically the same; when the PIR is greater than 0.01, the network latency of the topologies is greater than the other two topologies. The smaller the sub-net size, the better the network latency. When the network size is 8×8, that is, when the sub-net size is 2×2, there is a lower network latency. When the network scale is 16×16, it has a higher network latency due to the longer transmission distance.

Finally, from experimental results, 8x8 architecture has the lowest percentage of latency and dynamic energy consumption due to the dense distributions of 6 radio hubs in 64 cores WiNoC (the RH covered 37.5% of the entire 64 cores).

4.4. Effect of virtual channel on WiNoC

The experiments are based on three network sizes. The impact of various topology sizes (8x8, 12x12 and 16x16) used in this experiment with respect to VC and varying PIR have the same effect on different topology sizes, hence use the 8x8 architecture to demonstrate the rest experimented work.

4.5. Effect of virtual channel on network throughput:

Figure. 6 shows the network throughput comparisons for three VCs with respect to the 64 cores and varying PIR.When the PIR is less than 0.008, the throughput difference of the three virtual channels is very small; when the PIR reaches 0.008, 2VC reaches saturation; continue to increase the PIR, when the PIR reaches about 0.02, 4VC and 8VC reach saturation . 8VC has better performance at saturation point 0.02. The higher the VC, the higher the network throughput with respect varying PIR.

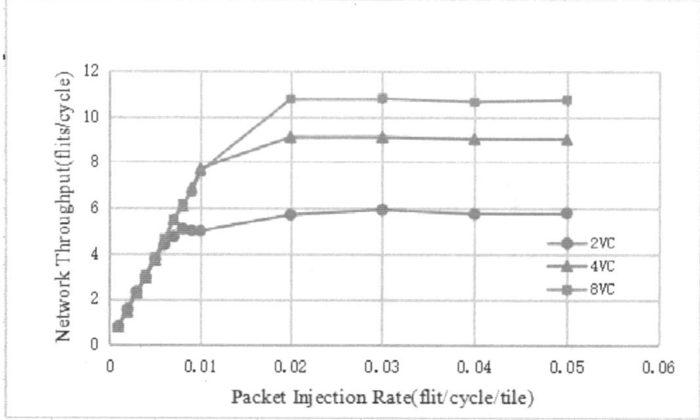

Figure 6. The network throughput for 64 cores

4.6. Effect of virtual channel on dynamic energy consumption:

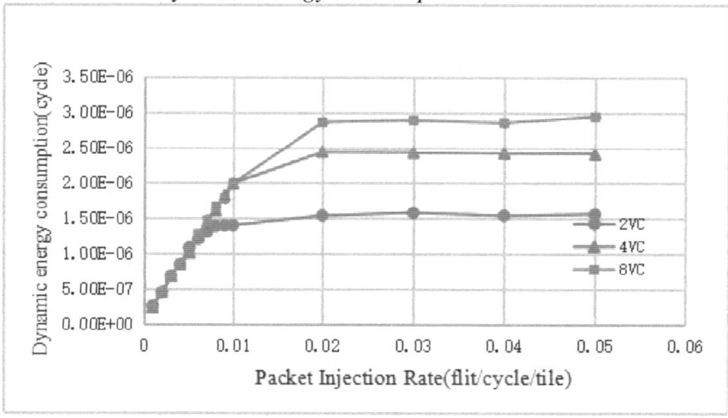

Figure 7. The dynamic energy for 64 cores

This section expatiates on the energy consumption comparisons of the VCs with respect to varying PIR. As can be observed from the experimental results tabulated in Figure 7, the energy consumption has an incremental trend with respect to increase in VCs. In the overall analysis, the 2VC have better energy consumption with respect to other VCs in the 64 cores. In WiNoC, the VC has a greater impact on energy consumption because the higher of VC with respect the network architecture sizes , the more energy consumption it consumes.

4.7. Effect of virtual channel on network latency:

It can be seen from the experimental results in Figure 8 that the network latency tends to increase with the increase of VCs. When the PIR is greater than 0.005, compared to 4VC and 8VC, the average delay of 2VC has a significant increase. This is because when the injection rate continues to increase, it takes a lot of time for the two virtual channels to process information.

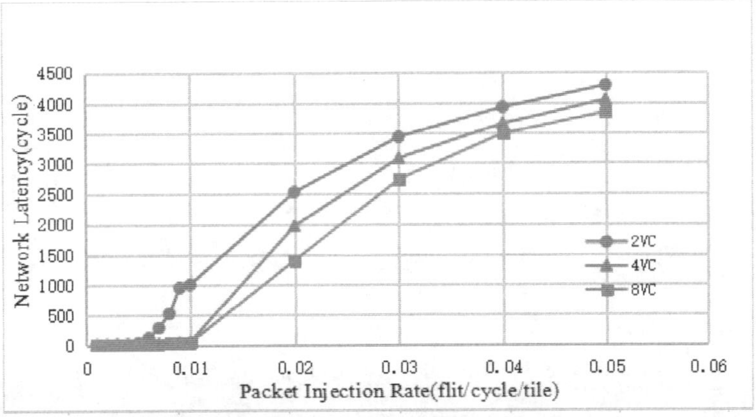

Figure 8. The network latency for 64 cores

5. Conclusion

The WiNoC can effectively solve the problem of the scalability limitation problem of the traditional wired network, and has a good development prospect. In this work, we proposed and evaluated a WiNoC architectures, using a co-evolution (CA) algorithm to determine the optimal number and optimal location of wireless nodes, and studied three different subnets under this architecture.The impact of scale on the use of WiNoC architecture. The number of RH has a great influence on the performance of the topology. Therefore, we should find an arrangement to minimize the number of radio hubs to obtain higher performance, scalability, high throughput and low latency. It provides a reference for the future research on hybrid WiNoC, how to select and how to distribute radio hubs in different situations, which will be helpful to the future research.

References

[1] Peng Y, Shigern Nakamura and J. Xu 2016 *10th 1EEEJACM International Symposium on Networks-on-Chip Conference* 10, 1109
[2] Tahanian E, Rezvani M and Fateh M *2021 26th International Computer Conference*
[3] Kim, Sung and Won 2016 *Journal of Sensors* 1, 27
[4] A. Carpenter, J. Hu, J. Xu, M. Huang and H. Wu 2011 *ACM SIGARCH Computer* 39, 271
[5] Maurizio P,Mario C and V Catania *2015 Journal of Low Power Electronics* 5, 38
[6] Dominic D,Avinash K and W. Rayess 2013
[7] Zhao D,Wang Y and Li J *2011 Proceedings of the Fifth Acm*, 177
[8] Ganguly A, Chang K and Deb S *2011 IEEE Transactions on Computers* 10, 1485
[9] Architectures Multicore Systems *2011 IEEE Transactions on Computers* 10, 1502
[10] DiTomaso D, Kodi A, Matolak D and W. Rayess *2014 IEEE Transactions on Parallel and Distributed Systems* 26, 3289
[11] Ganguly A, Chang K, Deb S and Teuscher C *2010 IEEE Transactions on Computers* 10, 1485
[12] Deb S,Chang K and Teuscher C *2012 13th International Symposium* 165
[13] *1997 Computers & Mathematics with Applications* 34, 136
[14] Potter M and Jong K *2000 Evol Comput* 8, 29
[15] Goh C and Tan K *2009 IEEE Trans. Evol. Comput* 13, 103
[16] Dijkstra E *Numerische Mathematik* 1959, 1 269
[17] Chen H,Wong K and Chung C *2006 IEEE Trans. Power Syst* 21, 143
[18] Catania V,Mineo A and Patti D *2015 IEEE 26th international conference* 162
[19] Fasiku A,Rusli S and Marsono M *2020 IEEE Student Conference on Research and Development* 27

The Design and Realization of Automobile Anti-theft Control Circuit Based on SCM

Xue Li*, Qisheng Li

Harbin Vocational & Technical College, Harbin, China

Corresponding author's email: 375317422@qq.com

Abstract. In this paper, the main realization of car theft and alarm, the system mainly to the thief can close to the car this behavior detection to explore and design, MCU as the central control processing system, constitute a complete system with data acquisition, object analysis, results upload, alarm trigger and other functions. Through the communication port of the system for data collection and transmission, can be the distance sensor, pyroelectric sensor and other recognition acquisition module for processing, and then through the alarm module, voice module for signal output.

1. Introduction

With the increasing of car ownership in our country, the anti-theft performance of the car has gradually attracted people's attention. In fact, as early as in foreign countries for the car anti-theft technology has long been studied. Because foreign car development especially in the fuel car development than our domestic earlier, also much faster. And foreign car theft cases also happen frequently, so in order to meet the needs of car users, many car companies are installed in the car anti-theft device, or have a special car anti-theft device for users to choose. With the progress of science and technology, car anti-theft device also from the beginning of the mechanical anti-theft lock, into now contain MCU processor anti-theft system[1-2].

With the increasing awareness of the protection of private territory and personal items, a lot of time and energy have been spent on the technical research of anti-theft, especially on the protection of valuables. Now most people have their own cars, for the car anti-theft is the modern people's incomparable concern.

2. Overall system design scheme

The system design framework is the minimum system of single chip microcomputer, which includes crystal oscillator circuit, single chip microcomputer circuit and reset circuit. Through the human body detection sensor and distance sensor and vibration sensor output signal input microcontroller detection, function keys control microcontroller operation, microcontroller output signal LCD display, and output to the alarm device[3-5].This paper mainly completes the software and hardware circuit design of car anti-theft system, including: control program, clock program, human body detection alarm, distance detection alarm, vibration detection alarm, key detection, data display, etc.

The whole system uses single chip microcomputer as the control core to realize the basic control functions of human body detection alarm, distance detection alarm, vibration detection alarm, control button detection, data display system. The main functions of the system include: data processing, start detection, to achieve the alarm alarm, signal transmission. The system design adopts the design concept of functional modularization. The system is mainly divided into three parts: overall scheme design, hardware design and software design. This system design module includes: power module, sensor module, display module, sound and light alarm module and MCU minimum system module. The schematic diagram of the overall scheme is shown in Figure 1.

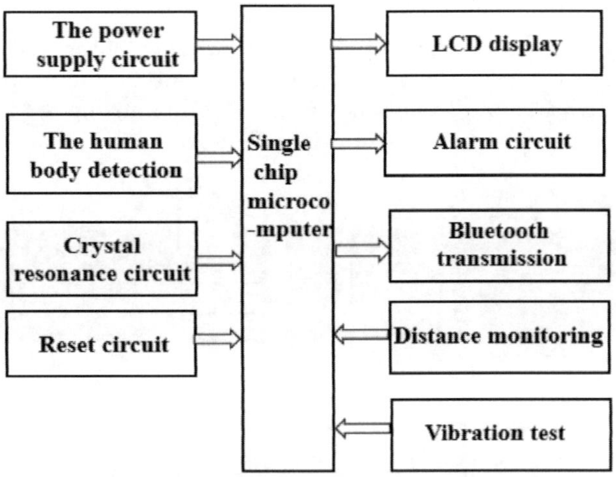

Figure 1. Schematic diagram of the overall scheme

3. System hardware design

3.1. STM32 microcontroller minimum system

STM32 minimum system contains crystal oscillator module, including power interface, reset system, debugging/download interface, start and other functions. The power module is 3.3V, and the capacitor of the power module mainly plays a role of filtering and isolating interference. The reset circuit is reset by level, so it is high level in normal operation. When the reset button is pressed, the reset circuit is grounded, and then the whole single-chip microcomputer system is reset.BOOT0 and BOOT1 pins of STM32 can be used to start the module from internal flash memory and system memory [6].The circuit diagram of the minimum system of STM32 MCU is shown in Figure 2.

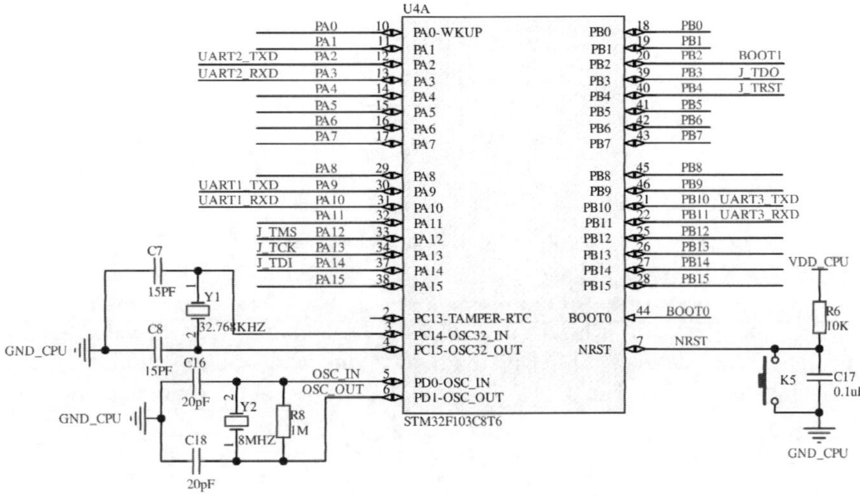

Figure 2. Microcontroller minimum system

Journal of Physics: Conference Series **2331** (2022) 012013 doi:10.1088/1742-6596/2331/1/012013

3.2. Distance sensor design

The range sensor uses an ultrasonic sensor to measure. Because according to the pin definition of ultrasonic module, 1 pin is connected to VDD to provide working voltage, 4 pin is connected to GND, 2 pin trigger control signal input, 3 pin echo signal input, through 2 pin input a high electric square wave, usually 20μs.After the module receives the high electric square wave, the module will automatically send 8 ultrasonic waves with a frequency of 40KHz. When the ultrasonic wave bounces back when encountering obstacles, it will be detected by the module [7].The module will pass 3 pin to MCU a high level signal. At this time, the SCM will record the time between triggering signal and receiving high level signal, and calculate the distance by multiplying the time by the sound speed of 340km/s. The distance detection circuit is shown in Figure 3.

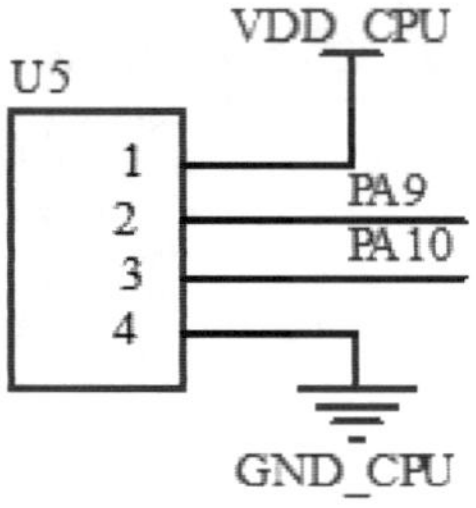

Figure 3. Distance detection circuit

3.3. Human body detection module design

According to the body temperature is 36.5°C, it will continue to emit 10μM infrared light. The human body detection module detects the infrared light emitted by the human body and focuses on the photosensitive infrared component driven by the Fresnel filter. After receiving the signal, it will lose the balance of cation and anion, releasing an electric charge, which is processed by a subsequent circuit to generate an alarm signal [8].Each pin function of the module: 1 pin is connected to 5V power supply, 2 pin is connected to output, and 3 pin is grounded. Human body detection circuit is shown in Figure 4.

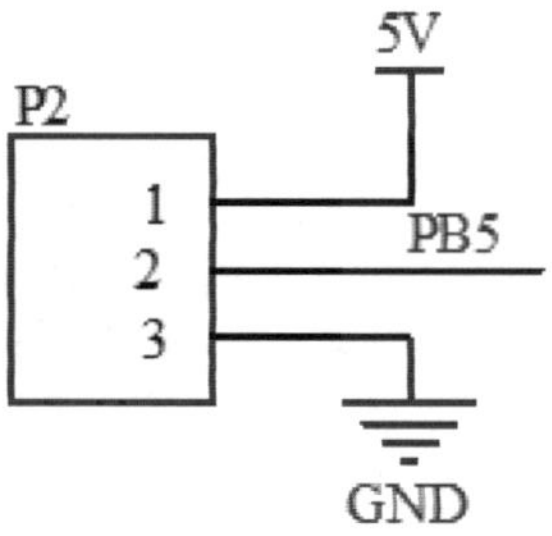

Figure 4. Human body detection circuit

3.4. Design of Bluetooth communication module

Embedded Bluetooth module is a kind of wireless data exchange between the whole system and mobile phones and other devices with Bluetooth receiving and sending modules [9].The Bluetooth module has six pins: STATE, R-T, T-R, GND, 5V and EN. Bluetooth communication circuit is shown in Figure 5.

Journal of Physics: Conference Series **2331** (2022) 012013 doi:10.1088/1742-6596/2331/1/012013

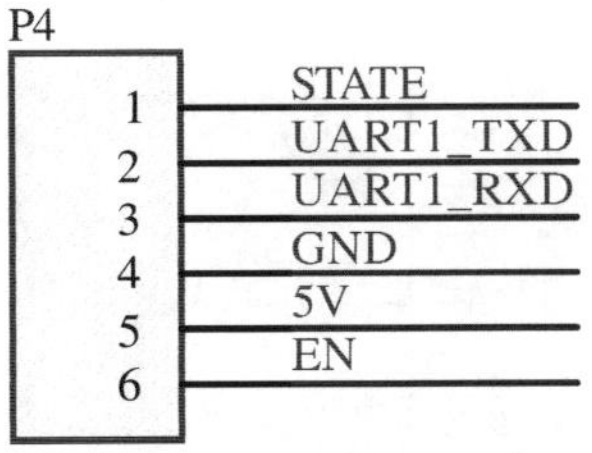

Figure 5. Bluetooth communication circuit

3.5. Design of vibration detection module

Using the module can trigger the control switch without directional abnormal vibration induction, any Angle can be triggered. The two pins of the vibration sensor are disconnected by default. When vibration occurs, the two pins are connected [10].

Under normal conditions, pin 3 of U2 comparator is pulled up to 3.3V by R1 resistance. According to the principle of series voltage division, the potential of pin 1 of comparator is equal to the voltage at both ends of R2.R2 and R3 are connected in series, and the voltage of comparator 1 pin is 1.65V, so the comparator outputs low level (logic 0).When the module is in vibration state, the two pins of the sensor are shorted. At this time, pin 3 of the comparator is directly grounded, so the comparator outputs high level (logic 1).Vibration detection module has three pins: 1 pin 5V, 2 pin output, 3 pin ground. Vibration detection circuit diagram is shown in Figure 6.

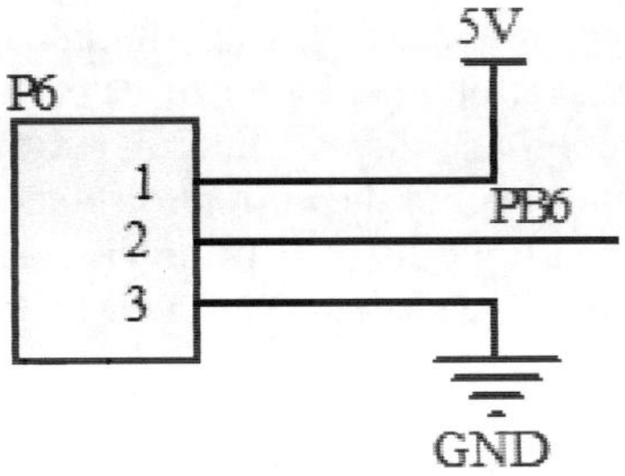

Figure 6. Vibration detection circuit

3.6. Display module design

Display module is the most commonly used LCD1206 module, which has a simple pin design. Programs are also easy to write. The data transmission pins of DB0~DB7 of LCD1206 are connected to PA0~PA7 of STM32.The other reset pins, read and write pins are also connected to PC13 to PC15 respectively. Backlight adjustment via VEE pins [11-12].The display circuit is shown in Figure 7.

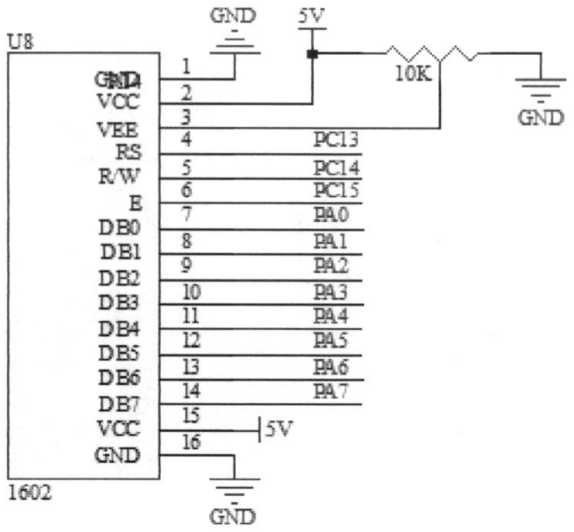

Figure 7. Shows the circuit

4. System software process design

The execution of the whole program needs to run the main program first, subroutines are generally called by the main program, so as to achieve various functions, that is, the initialization of the software system, the realization of data display, processing, parameters, parameters transfer a series of functions [13-14].The main program also plays a decisive role in the execution timing of each subroutine. Software system subroutine module includes: LCD subroutine, alarm subroutine, keystroke adjustment subroutine and so on. After first began to enter the system initialization, such as initialization buttons, LED display screen, the system outage, etc., and then to detect and display the distance, vibration detection and display, have no detection and display, interrupted by two buttons to set the exceed the detection distance, if the judge did not exceed the current distance distance set, continued to test, the buzzer alarm, exceeding the maximum limit distance If the vibration limit is exceeded, the alarm will be sent out; if the vibration limit is not exceeded, the alarm will be sent out. The program flow chart is shown in Figure 8.

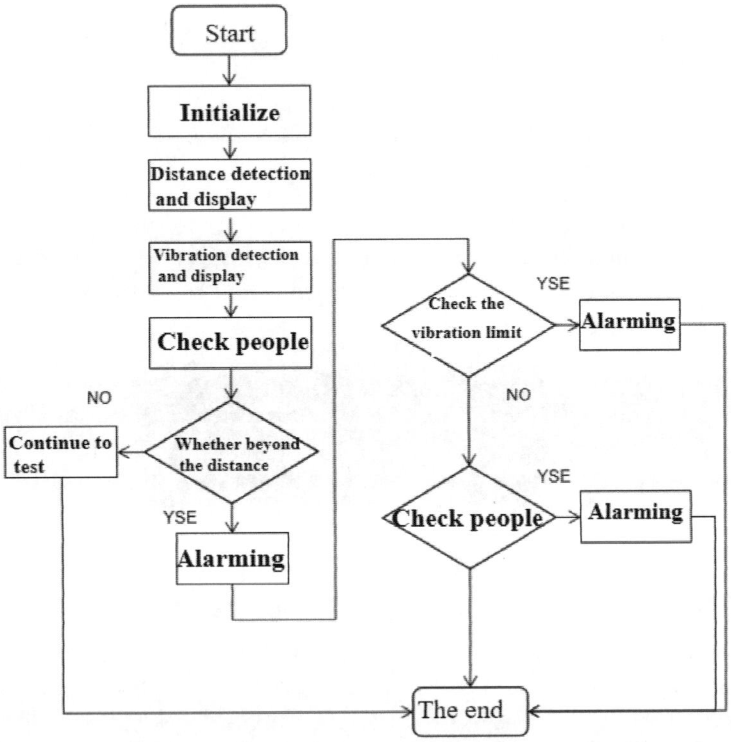

Figure 8. Shows the program flow chart

5. System debugging and testing

5.1 Hardware Debugging

Because the design involves more modules, the system includes sensor module, SCM module, display alarm, so it is difficult to debug, the design of uncertain factors are also more, so the use of block debugging method, eliminate the interference of each module. After completing the above checks and confirming no errors, power debugging can be carried out, but at this time, attention should be paid to the positive and negative polarity of the power supply cannot be connected. The physical picture of automobile anti-theft system is shown in Figure 9.

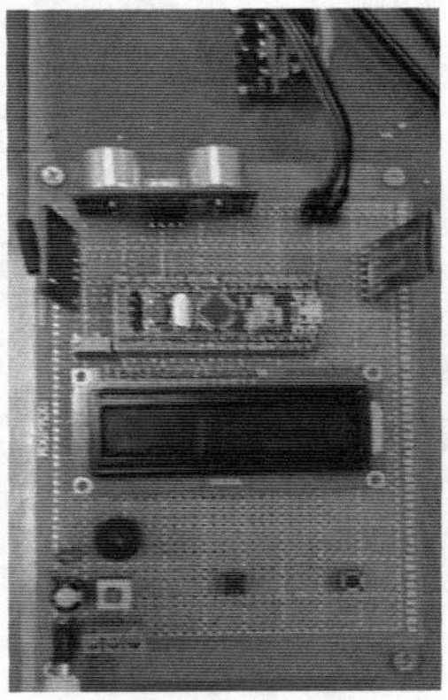

Figure 9. Physical picture of automobile anti-theft system

There are four buttons on the app interface of the mobile phone, namely, Turn on Bluetooth, connect Bluetooth, turn off Bluetooth and close the program. Can display distance, body detection, vibration detection results.

5.2. Screen Test

Screen display is LCD1206, this display module is widely used. Because of its simple pin, the program is easy to use. According to the program Settings, after powering on, each module will detect and display the detection data on the display screen, and you can feel whether the displayed data is correct through actual measurement. If the screen is too bright not to show, then it should be adjusted to display circuit potentiometer, because this problem is the display backlight is not adjusted. If the display is not bright, it should be the power supply of the hardware circuit is unstable. The screen should look like Figure 10.

Figure 10. Screen test

5.3. Distance Detection

Distance detection using ultrasonic sensor scheme, in order to test the accuracy of distance in the program.STM32 will test the distance 5 times in a row to take the average and display the average on the screen. The Dis on the screen shows the distance. During the test, it should be noted that the ultrasonic module cannot test too short a distance, because the interval time of signal recognition is too short for the system to calculate the actual distance [15].

Here, the speed of sound is generally selected as 340m/s, but the measured speed of sound may be less than 340m/s, resulting in inaccurate distance. At this time, a ruler can be used to mark several accurate distances as the speed of sound in the debugging program, so as to obtain an accurate distance test. The distance module test is shown in Figure 11.

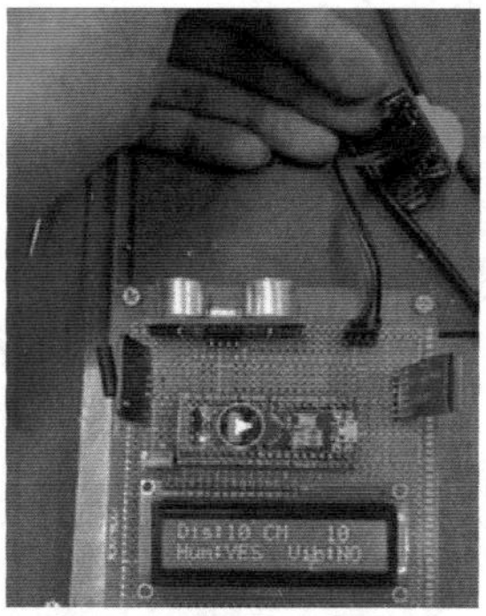

Figure 11. Distance module test

5.4. Human detection

The human body detection module recognizes the specific infrared ray released by human body, and then gives STM32 a high level to indicate the detection of human body, and then displays Hum: yes on the display board. The physical test of human body detection module is shown in Figure 12.

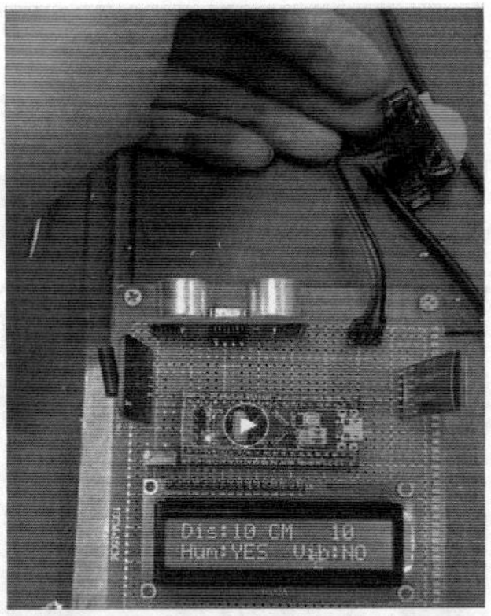

Figure 12. Human body detection module test

5.5. Vibration Detection

The vibration detection module detects whether the object is vibrating. If vibration occurs, it will give STM32 a high level to detect vibration, and then display Vib: yes on the display board. Physical test of vibration detection module is shown in Figure 13.

Figure 13. Vibration detection module test

5.6. Alarm module detection

Whenever any of the above modules detects a value that exceeds the set value, an alarm circuit is required to alert the user of a problem. The system is operated by buzzer and light-emitting diode. The actual measurement of alarm circuit is shown in Figure 14.

Figure 14. Alarm module test

5.7. Experimental Results
Through the detailed functional test of the hardware, software, functional modules and mobile app of this design, according to the test, the actual function of this design is completely consistent with the design function, which can achieve the expected effect. The test results are shown in Table 1. It is a very good and innovative design.

Table 1. Test statistics

The serial number	Test module	A plan	The actual function	The test results
1	Display module	Initialization display	Meet the requirements	normal
2	Human body detection module	Test to see if anyone passes	Meet the requirements	normal
3	Ultrasonic ranging module	Make distance measurement	Meet the requirements	normal
4	The vibration module	Check whether the car vibrates	Meet the requirements	normal
5	Alarm module	Alarm when the set value is exceeded	Meet the requirements	normal
6	The wireless module	To transmit information	Meet the requirements	normal

6. Conclusion
This topic about car alarm control circuit was introduced in detail the design and implementation, although there has been a related automobile anti-theft technology on the market, but due to the large volume, high cost, hindered the purchase desire of most consumers, therefore, the development of the design, on the one hand, the development of single-chip microcomputer in the future, the extension of different areas and so on, is the development of the important significance, At the same time, it is of great auxiliary significance to the automobile industry.

The hardware equipment is applied to the STM32 minimum system board, ultrasonic sensor, human body detection sensor, vibration detection module, display module, alarm module, Bluetooth module and so on. In the use of software, circuit diagram drawing application Altium Designer, program compilation application Keil uVision4.Moreover, the mobile phone APP programming software was innovatively applied, and the Bluetooth app on the mobile phone was designed in a graphical way.

According to the design requirements, mainly through the MCU control, to achieve someone close, automatic alarm function, at the same time close to the distance information display, and when the car vibration, will also send an alarm to prevent theft, in many aspects of the overall car anti-theft system design. After debugging and testing, the whole system can meet the requirements of the design task and work well.

Aiming at the disadvantages of the existing automobile anti-theft products, such as single anti-theft function, low reliability and lack of tracking and monitoring function, an intelligent automobile anti-theft alarm system is designed. The whole system uses single-chip microcomputer as the control core, the way of combining hardware and software system, to realize the human body detection alarm, distance detection alarm, vibration detection alarm, control button detection, data display system and other basic control functions. This design can undertake automobile vibration alarm function, also can realize the function and alarm at the same time, the use of sound and light alarm system to reflect, using distance sensors can also detect the suspects and the distance of the car, if necessary can add wireless bluetooth module sends the owners, etc., has a fully functional, simple production, low cost, high cost performance.

7. Reference

[1] Chen Chengyi, Peng Peng, Chen Xiaodong. Application of MBR process in wastewater treatment of printed circuit board. Construction Science and Technology, 2017(21):84-85.

[2] Yang Lingling, WANG Yang. Exploration of Project-based Integration Teaching Mode of Electronic Technology foundation and PCB Design Course. Education modernization, 2019, 6(97):142-144.

[3] Zhang Xiaodong, Liu Siheng, LIAO Jianjun. Research on teaching Reform of PCB Circuit Design Course. Education modernization, 2018, 5(01):73-74+101.

[4] Yang Hongqiang. Discussion on the past, present and future of printed circuit board. Printed circuit information, 2018, 26(01):6-15.

[5] Zhang Cheng, You Yongfeng. Design principle and Fabrication method of PCB by PROTEL DXP in Laboratory. Journal of Yangling Vocational and Technical College, 2007(04):45-46.

[6] Zhang Kaiheng. What preparation should be made before PCB design. Electronic Journal, 2020-01-12(006).

[7] Zhang Chunxiang, Wang Min. Printed circuit board manufacturing skills. Jiangxi Chemical Industry, 2017(06):7-10.

[8] Li TONG. Application of polytetrafluoroethylene emulsion in high frequency copper cladding plate. Organic Fluorine Industry, 2019(01):23-27.

[9] Huang Liang, CUI Xinyue. Steps of PCB fabrication in laboratory and Problems needing attention. Neijiang Technology, 2008, 29(12):173.

[10] Li peng. Design method of PCB board and its application in experimental teaching. Education modernization, 2017, 4(16):61-62.

[11] Qi Zhongqi, LIU Chuxiang. Method of Making PCB by hand and Analysis of Electrical Characteristics of product. China Education Technology Equipment, 2011(30):103-105.

[12] LIU Houchuan, QI Jianjian, Lv ZhaohuiResearch progress in the treatment of waste liquid of printed circuit board etching and copper etching. Contemporary Chemical Research, 2019(15):109-110.

[13] Fan Qiang, DONG Hongbo, Ma Bin. Design of temperature control system based on PLC and touch screen. Mechanical research & application, 2019, 32(03):154-155+158.

[14] Lv Zhihua. Design of Temperature Control System based on SCM. Electronic Technology and Software Engineering, 2019(21):231-232.

[15] Fan Qiang, DONG Hongbo, Ma Bin. Design of temperature control system based on PLC and touch screen. Mechanical research & application, 2019, 32(03):154-155+158.

Journal of Physics: Conference Series **2331** (2022) 012014 doi:10.1088/1742-6596/2331/1/012014

Prediction Model of Elastic Load Resource Regulation Potential under Hierarchical and Partitioned Dynamic Control Architecture

Ying ZHOU[1,*], Songsong CHEN[1], Kun SHI[1], Taorong GONG[1], Jindou YUAN[1], Bo LI[2] and Shihai YANG[2]

[1]China Electric Power Research Institute Co., Ltd.: Beijing Key Laboratory of Demand Side Multi-Energy Carriers Optimization and Interaction Technique, Beijing 100192, China
[2]State Grid Jiangsu Electric Power Co., Ltd, Nanjing 210000, China

Corresponding author's e-mail: hgzhouying@163.com

Abstract. In order to deal with the problem of energy shortage, demand side management has attracted more and more attention. Taking demand side resources as an example to understand the regulation potential of demand side resources, this paper designs a hierarchical and partitioned dynamic regulation architecture for demand side adjustable resources, which can carry out the interconnection between multi-level subjects and efficient information interaction. The regulation potential prediction model of elastic load is established through the hierarchical and partitioned architecture, and the credible and adjustable potential of resident users under two regulation scenarios is obtained through calculation: participating in emergency load reduction and peak valley difference, which proves the effectiveness of the architecture.

1. Introduction

With the rapid development of the energy Internet era, the energy demand of users shows explosive growth, and there is a power gap during peak hours. There are too many base load power sources on the power side and less flexible resources, especially peak resources, so the peak load demand can not be met[1]. Through the assessment of adjustable load regulation capacity and the implementation of demand side management, the capital cost can be greatly reduced and this problem can be solved efficiently[2],[3].

At present, foreign countries have basically formed a set of schemes for the evaluation of users' adjustable ability. At present, foreign countries have basically formed a set of schemes for the evaluation of users' adjustable ability. The DOE's evaluation of the elastic load response regulation capacity in the national power market lists the following three schemes: first, the example analysis, based on the prediction of the user's elastic load usage before the implementation of the response task, emphasizes the regulation potential and economic benefits of the elastic load response, and evaluates the results[4]; Second, comprehensive resource planning, based on avoiding losses caused by users' elastic load in response to projects and affecting their own power consumption, evaluates the available resources for users' elastic load to participate in the response, which is suitable for long-term evaluation in an environment without retail competition and vertical management of public utilities[5]; Third, case performance analysis, based on the actual project and measured data, compares and analyses with the expected objectives, and adopts the concept of standardized total benefit[6], which is convenient to compare the implementation effects of different load response projects[7].

However, considering that different regions and different types of load resources may show great differences in different environmental states, and the same type of load may also show obvious differences at different times, the current potential analysis method can not achieve such refined information collection and potential evaluation; The lack of this information will be difficult to describe the adjustable interaction behaviour of demand side resources, and then can not accurately and comprehensively evaluate the power efficiency level and energy consumption characteristics of demand side. Based on this, this paper proposes a hierarchical and partitioned flexibility architecture, through which regional fine information perception can be carried out, and many useful information contained in the demand side can be deeply excavated through the efficient information processing ability between levels. Aggregation and cross regional aggregation can be carried out according to different regulation requirements and regulation scenarios, so as to realize the accurate potential evaluation of load polymers.

The structure of this paper is as follows: section II introduces the flexible architecture of hierarchical and partitioned dynamic control. Section III introduces the physical model of adjustable load and the calculation model of adjustable potential. Section IV calculates the load adjustable potential under different regulation scenarios. Section V summarizes this paper.

2. Hierarchical and partitioned dynamic control architecture

The hierarchical and partitioned dynamic control architecture includes two parts: layering and zoning. Partition refers to regional aggregation of loads according to geographical location according to load distribution and load characteristics. At the same time, it can also carry out cross regional virtual aggregation of adjacent regions through information exchange among multiple regions; Hierarchical architecture refers to the scheduling and control of controllable loads through multi-level horizontal communication and vertical information transmission according to the power grid management level, including four levels: elastic load operation platform, load aggregator, edge agent and terminal load. The specific model is shown in the Figure 1.

Through the flexible hierarchical and partitioned architecture, on the premise of ensuring the controllability of the load system, the collaborative adjustment of multiple types of loads is realized through the schedulable capacity allocation of the cloud and the edge. Through the management of the edge IOT agent, the similar loads installed in the same regional power grid are aggregated with lower delay, less bandwidth and computing resources, and controlled through a unified aggregation model and control instructions to form an effective load transfer capability. On this basis, the adjustable potential analysis of load polymer will effectively improve the accuracy of potential prediction.

At present, there is no unified definition of elastic load in domestic and foreign research. It is generally considered that elastic load is summarized as a load whose size can be expanded and retracted within the interval specified by the user[8]. Considering the friendly interaction between power grid and user supply and demand, this paper defines elastic load from demand side and supply side.

1) On the demand side, the elastic load is defined as the load that can actively participate in the power grid regulation and dynamically adjust and actively respond when the supply-demand balance of the power grid changes;

2) On the supply side, the elastic load is defined as the load that the power grid can monitor and accurately perceive its power consumption nature, power consumption law and power consumption state in real time, and can realize transfer, interruption and selective removal through load regulation technology within the regulation time and regulation area set by the power grid, so as to realize flexible regulation.

Under the hierarchical and partitioned architecture, the adjustable potential of elastic load can be accurately analyzed through centralized distributed control. The elastic load control platform in the cloud is responsible for the centralized control and prediction information of load clusters, including the adjustable potential prediction of elastic load, the output prediction of distributed power generation, etc. According to the forecast information, the regulation tasks can be assigned to the load aggregators at all levels at the lower level. After accepting the tasks, the load aggregators reasonably arrange the plans for various loads to participate in the regulation tasks, and send the instructions to the edge IOT agent, so as to realize the load regulation and information collection through the edge IOT agent. The process of aggregator regulating the load in the jurisdiction is called distributed control. In the process of potential prediction, the edge IOT agent is mainly responsible for the collection of load information and environmental information. The aggregator aggregates the load, and the management and control platform forecasts the potential according to the polymer model.

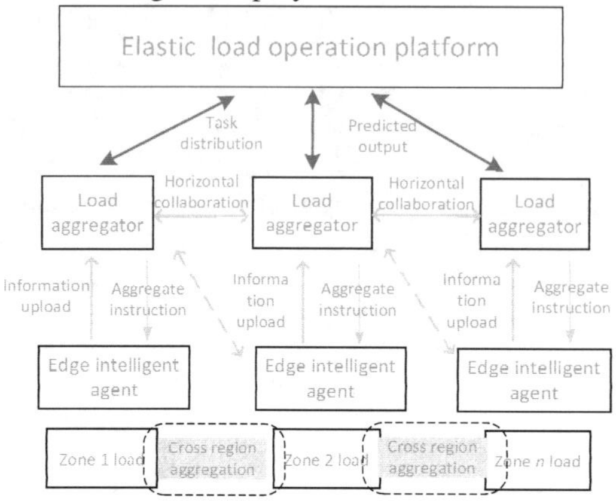

Figure 1. Hierarchical and partitioned dynamic control architecture

3. Physical model construction

3.1. Adjustable potential of air conditioning load

When the air conditioner is in the refrigeration mode, its actual power consumption is related to the working state of the air conditioner. When the room temperature is higher than the maximum value, the air conditioner is in the power on refrigeration state; When the room temperature is lower than the minimum value, the air conditioner is in the power-off state, and its power consumption is almost 0; When the room temperature is within the temperature setting range of the air conditioner, the air conditioner remains in its original state. The actual power of air conditioning in t period is calculated as follows:

$$p_{AC,t} = P_{AC} \times S_{AC,t} \tag{1}$$

$$S_{AC,t+1} = \begin{cases} 0, T_{AC,t+1} \le T_{AC\min} \\ 1, T_{AC,t+1} \ge T_{AC\max} \le T_{AC\min} \\ S_{AC,t}, T_{AC\min} < T_{AC,t+1} < T_{AC\max} \end{cases} \tag{2}$$

Where, $p_{AC,t}$ is the actual power of air conditioning during the t period; P_{AC} is the rated power of air conditioner under refrigeration; $S_{AC,t}$ is the working state of the air conditioner during the t period (a value of 0 means power off; a value of 1 means power on); $T_{AC\min}$ is the minimum room temperature setting; ΔT_{AC} set the range for room temperature; $T_{AC,t}$ is the room temperature of the t period.

In the refrigeration mode, the physical model of the air conditioner, i.e. the room temperature in the output variable period, is as follows:

$$T_{AC,t+1} = T_{AC,t} + \Delta t \cdot \frac{G_t}{\Delta c} + \Delta t \cdot \frac{C_{AC}}{\Delta c} \cdot S_{AC,t} \tag{3}$$

Where, $T_{AC,t+1}$ and $T_{AC,t}$ respectively represent the $t+1$ period and the t period room temperature during the time period; G_t represents the outdoor and indoor heat exchange value during the t period, and Δc represents the indoor temperature coefficient, that is, the heat required for every 10 ℃ increase in room temperature; C_{AC} indicates the cooling capacity under the condition of heat; $C_{AC} / \Delta c$ indicates the action value of the air conditioner on the change of room temperature under refrigeration; Δt represents the time interval, which is taken as 1min.

For the convenience of calculation, after linearizing the above variables, the physical model of air conditioning is expressed as follows:

$$T_{AC,t+1} = T_{AC,t} + 0.4 - 0.8S_{AC,t} \tag{4}$$

when $S_{AC,t} = 1$, $T_{AC,t+1} = T_{AC,t} + 0.4 - 0.8S_{AC,t}$; when $S_{AC,t} = 0$, $T_{AC,t+1} = T_{AC,t} + 0.4$.

The calculation formula of adjustable potential of air conditioner can be obtained:

$$P^u_{\text{pot, }AC} = S_{AC}(t)P^u_{\text{pot, }AC} \tag{5}$$

Calculation of minimum adjustable physical potential:

$$P^d_{\text{pot, }AC} = S_{AC}(t)P^d_{\text{pot, }AC} \tag{6}$$

According to the response reliability evaluation method, the equipment credible response degree r_{AC-cre}, combined with the adjustable potential, the air conditioning credible adjustable potential is calculated.

$$P_{cf,AC} = \frac{(P^d_{pot,AC} + P^u_{pot,AC}) \bullet r_{AC-cre}}{2} \tag{7}$$

The credible adjustable potential of the air conditioning equipment has been calculated above, so the confidence interval of the adjustment potential under the confidence level L is $\left[P^d_{cf,AC}, P^u_{cf,AC}\right]$。

$$P^d_{cf,AC} = P^d_{\text{pot, }AC} + (P_{cf,AC} - P^d_{\text{pot, }AC})(1-L)^{0.5} \tag{8}$$

$$P^u_{cf,AC} = P^u_{\text{pot, }AC} + (P^u_{\text{pot, }AC} - P_{cf,AC})(1-L)^{0.5} \tag{9}$$

Where, $P_{cf,AC}$ is the most credible regulatory potential, $P^u_{cf,AC}$ is the upper limit of credible adjustable potential, $P^d_{cf,AC}$ is the lower limit of credible adjustable potential.

3.2. Adjustable potential of resident users

For residential users, according to different equipment types, the responsive equipment of residential users generally includes air conditioner, water heater, refrigerator, electric vehicle, etc. the calculation formula of adjustable potential of residential users is:

$$P^u_{\text{pot,res}} = \sum_{i=1}^{n_1}\sum_{j=1}^{n_2} S_{res,i,j}(t)P^u_{\text{pot,res,i,j}} \tag{10}$$

Calculation of minimum adjustable physical potential:

$$P^d_{\text{pot,}res} = \sum_{i=1}^{n_1}\sum_{j=1}^{n_2} S_{res,i,j}(t)P^d_{\text{pot,res,i,j}} \tag{11}$$

According to the user's response reliability evaluation method $r_{res-cre}$, after iterative calculation, the multi-user trusted response degree can be obtained. Combined with the adjustable potential, the user trusted adjustable potential can be calculated.

$$P_{cf,res} = \frac{(P^d_{pot,res} + P^u_{pot,res}) \bullet r_{res-cre}}{2} \tag{12}$$

The maximum physical regulation potential and the most credible regulation potential that the resident user can respond to the load have been calculated above, so the confidence interval of the regulation potential under the confidence level L is $\left[P^d_{cf,res}, P^u_{cf,res}\right]$。

$$P^d_{cf,res} = P^d_{\text{pot, res}} + (P_{cf,res} - P^d_{\text{pot,res}})(1-L)^{0.5} \tag{13}$$

$$P^u_{cf,res} = P^u_{\text{pot,res}} + (P^u_{\text{pot,res}} - P_{cf,res})(1-L)^{0.5} \tag{14}$$

Where, $P_{cf,res}$ is the most credible regulatory potential, $P^u_{cf,res}$ is the upper limit of credible adjustable potential, $P^d_{cf,res}$ is the lower limit of credible adjustable potential.

For commercial users, the adjustable equipment is generally central air conditioning, electric vehicles, etc. the calculation method of the credible adjustable potential and confidence interval of commercial users is the same as that of commercial users.

4. Simulation results and analysis

4.1. Resident users participate in emergency load reduction

According to different market rules of demand response, set aggregate users to participate in the following emergency demand response scenarios, and calculate the total credible and adjustable potential of 2885 resident users in the demonstration area to participate in the emergency demand response scenario within 24 hours, as shown in Figure 2.

From the analysis of Figure 2, It can be seen from the figure that the adjustable potential of resident users in the emergency demand response scenario is the largest, followed by electric vehicles, and the adjustable potential of temperature control load in the emergency demand response scenario is the smallest. And the adjustable potential of the three increased significantly from 10 a.m. to 21 p.m. Taking the user at 15:00 p.m. as an example, the total adjustable power of residential load is 4269.768kW, of which the adjustable potential of electric vehicle is 3183.7kW and the adjustable potential of temperature control load is 1043.86kW.

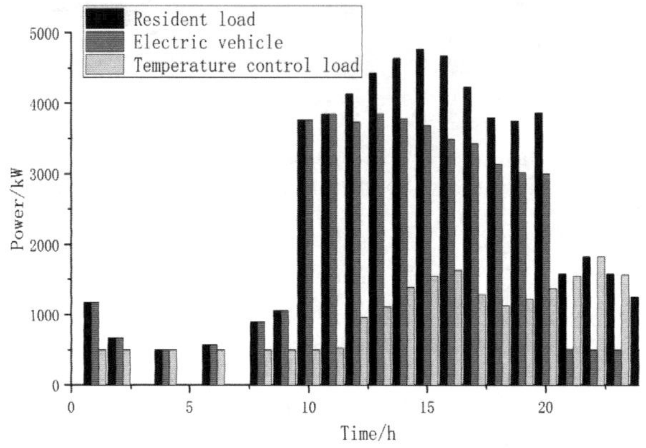

Figure 2. Prediction curve of various loads

5. Conclusion

This paper introduces a hierarchical and partitioned dynamic control architecture, and realizes the regulation potential prediction of elastic load based on the multi-level interconnection and information exchange of this architecture. Then, through the calculation of the model, the credible and adjustable potential of resident users under the two control scenarios of emergency load reduction and peak valley difference reduction is obtained.

Acknowledgments

Thanks for the project supported by the hierarchical and partitioned adjustable load aggregation and active interactive control technology (5100-202118394A-0-0-00) and the support of the science and technology project managed by the headquarters of the State Grid Corporation of China.

References

[1] Li HENGQIANG, Wang Jun, Jiang XINGRAN, et al Air conditioning load regulation strategy for power grid peak shaving auxiliary service and its virtual unit modeling [J] Power construction, 2020, 41 (7): 9
[2] Zhu Yuxiang, Zhang Junfang, Cheng Weijie, et al Risk assessment of urban power grid based on demand side management [J] Modern power, 2020, 37 (2): 6
[3] Lu Jie, Liu Xiang, Li Haoxiang, Tian Jing Research on load demand response potential of household appliances based on SOM neural network [J] Machinery and electronics, 2020,38 (11): 47-49 + 54.
[4] Mansouri M R , Fuchs P F . Elastic load coupling with tailored elastomer composites[J]. 2021.
[5] Shen Z Y . Risk Analysis and Measures for Deep Peak Load Regulation of Ultra Supercritical 1000MW Unit[J]. Energy conservation and environmental protection, 2019
[6] Svanadze M . Potential method in the coupled theory of elastic double-porosity materials[J]. Acta Mechanica, 2021, 232(6):2307-2329.

[7] Liu meizhao, Zhang Dongyu, Luo Hui Line potential evaluation based on load curve decomposition [C] / / Proceedings of the third smart grid Conference - smart power, 2019:116-124
[8] Jiang Dongrong, Xiang Hongwei, Wang Yiwen, et al Research on incentive pricing strategy of load aggregator and controllable load scheduling based on master-slave game [J] Journal of Chongqing University of Technology: Natural Science, 2018, 32 (5): 8

AUTHOR INDEX

Chen, Songsong 106
Chen, Tianrui 38
Gong, Taorong 106
Hongbin, Pu 10
Jianning, Xu 10
Jiao, Zhi-Wei 72
Jichao, Hu 10
Li, Bo 106
Li, Qisheng 96
Li, Wenliang 61
Li, Xue 96
Liu, Liang 81
Luo, Bin 72
Na, Li 10
Sangwaranatee, N 51
Sangwaranatee, Narong 56
Shen, Sili 24
Shi, Kun 106
Song, Chaoyang 1
Song, Guozhi 87
Srithanachai, I 51
Srithanachai, Itsara 56
Wang, Bohan 1
Wang, Dian 81
Wang, Feipeng 15
Wang, Hao-Yu 72
Wang, Huimin 24
Wu, Xingwei 81
Wu, Zhihao 24
Xi, Wang 10
Xin, Yin 87
Yang, Shihai 106
Yang, Wei-Min 72
Yin, Yue 81
Ying, Zhijun 24
Yingxiang, Yang 10
Yu, Yuan 72
Yuan, Jindou 106
Zhang, Yingying 46
Zhang, Zheng 15
Zhao, Yunfei 24
Zhou, Ying 106